Dominik Schleicher

The early universe

Dominik Schleicher

The early universe

Probing primordial magnetic fields, dark matter models and the first supermassive black holes

Südwestdeutscher Verlag für Hochschulschriften

Impressum/Imprint (nur für Deutschland/ only for Germany)
Bibliografische Information der Deutschen Nationalbibliothek: Die Deutsche Nationalbibliothek verzeichnet diese Publikation in der Deutschen Nationalbibliografie; detaillierte bibliografische Daten sind im Internet über http://dnb.d-nb.de abrufbar.

Alle in diesem Buch genannten Marken und Produktnamen unterliegen warenzeichen-, marken- oder patentrechtlichem Schutz bzw. sind Warenzeichen oder eingetragene Warenzeichen der jeweiligen Inhaber. Die Wiedergabe von Marken, Produktnamen, Gebrauchsnamen, Handelsnamen, Warenbezeichnungen u.s.w. in diesem Werk berechtigt auch ohne besondere Kennzeichnung nicht zu der Annahme, dass solche Namen im Sinne der Warenzeichen- und Markenschutzgesetzgebung als frei zu betrachten wären und daher von jedermann benutzt werden dürften.

Verlag: Südwestdeutscher Verlag für Hochschulschriften Aktiengesellschaft & Co. KG
Dudweiler Landstr. 99, 66123 Saarbrücken, Deutschland
Telefon +49 681 37 20 271-1, Telefax +49 681 37 20 271-0
Email: info@svh-verlag.de
Zugl.: Heidelberg, Ruprecht-Karls-Universität, Diss., 2009 (überarbeitete Version)

Herstellung in Deutschland:
Schaltungsdienst Lange o.H.G., Berlin
Books on Demand GmbH, Norderstedt
Reha GmbH, Saarbrücken
Amazon Distribution GmbH, Leipzig
ISBN: 978-3-8381-1785-0

Imprint (only for USA, GB)
Bibliographic information published by the Deutsche Nationalbibliothek: The Deutsche Nationalbibliothek lists this publication in the Deutsche Nationalbibliografie; detailed bibliographic data are available in the Internet at http://dnb.d-nb.de.

Any brand names and product names mentioned in this book are subject to trademark, brand or patent protection and are trademarks or registered trademarks of their respective holders. The use of brand names, product names, common names, trade names, product descriptions etc. even without a particular marking in this works is in no way to be construed to mean that such names may be regarded as unrestricted in respect of trademark and brand protection legislation and could thus be used by anyone.

Publisher: Südwestdeutscher Verlag für Hochschulschriften Aktiengesellschaft & Co. KG
Dudweiler Landstr. 99, 66123 Saarbrücken, Germany
Phone +49 681 37 20 271-1, Fax +49 681 37 20 271-0
Email: info@svh-verlag.de

Printed in the U.S.A.
Printed in the U.K. by (see last page)
ISBN: 978-3-8381-1785-0

Copyright © 2010 by the author and Südwestdeutscher Verlag für Hochschulschriften Aktiengesellschaft & Co. KG and licensors
All rights reserved. Saarbrücken 2010

The early universe:
Probing primordial magnetic fields, dark matter models and the first supermassive black holes

Referees: Prof. Dr. Ralf S. Klessen
                       Prof. Dr. Matthias Bartelmann

DISSERTATION
SUBMITTED TO THE
COMBINED FACULTIES OF THE NATURAL SCIENCES AND MATHEMATICS
OF THE RUPERTO-CAROLA-UNIVERSITY OF HEIDELBERG, GERMANY
FOR THE DEGREE OF
DOCTOR OF NATURAL SCIENCES

Updated version (May 2010)

PUT FORWARD BY

DIPL. PHYS. DOMINIK R. G. SCHLEICHER
BORN IN: BAMBERG (GERMANY)

ORAL EXAMINATION: JUNE 9th, 2009

Für meine Eltern

Zusammenfassung

Das Ziel dieser Arbeit ist es, das frühe Universum zwischen Rekombination und Reionisation besser zu verstehen und neue Möglichkeiten aufzuzeigen, es zu erforschen. Dies betrifft die Sternpopulation, die Physik im frühen Universum und die Entstehung der ersten supermassereichen Schwarzen Löcher. Mit Hilfe der von WMAP 5 gemessene optische Tiefe der Reionisation wurden obere Schranken für die Stärke von primordialen Magnetfeldern sowie die Annihilation und den Zerfall dunkler Materie und Einschränkungen für Modelle der Sternpopulation abgeleitet. Durch den Gammastrahlungs- und Neutrinohintergrund lassen sich weitere Modelleinschränkungen ableiten, die besonders für leichte dunkle Materie signifikant sind. Ferner wurde gezeigt, dass zukünftige 21 cm Beobachtungen wesentlich stärkere Einschränkungen an primordiale Magnetfeldstärken liefern. Um den Ursprung der ersten supermassereichen Schwarzen Löcher und ihrer hohen Metallizität besser zu verstehen, wurde untersucht, wie sich diese mit ALMA zwischen Rotverschiebung 5 und 15 beobachten lassen. Hierfür wurden verschiedene Beobachtungsgrößen abgeschätzt und klassifiziert, und die Anzahl der verfügbaren Quellen abgeschätzt. Trotz großer Unsicherheiten ist zu erwarten, dass einige Quellen in einem Raumwinkel entsprechend dem Hubble-Deep-Field gefunden werden können.

Abstract

The goal of this work is to better understand the universe between recombination and reionization and to outline new possibilities to explore it in more detail. This concerns the stellar population, the physics of the early universe, and the formation of the first supermassive black holes. With the reionization optical depth from WMAP 5, I derive upper limits for the strength of primordial magnetic fields and dark matter annihilation / decay, as well as constraints for stellar population models. Further constraints can be found from the gamma-ray and neutrino background, which rule out s-wave annihilation of light dark matter. It was shown that future 21 cm observations will constrain primordial magnetic fields even further. To improve our understanding of the origin of the first supermassive black holes and their high metallicity, I explore how they can be observed with ALMA between redshift 5 and 15. For this purpose, I estimated and classified the available observables, and I provided independent estimates for the expected number of high-redshift black holes. In spite of large model uncertainties, one can expect to find at least a few sources in a solid angle similar to the Hubble-Deep-Field.

Contents

List of Figures   ix

List of Tables   xvii

1 Introduction   1
   1.1 The timeline of cosmology 1
      1.1.1 From the Big Bang to cosmic recombination 2
      1.1.2 From the postrecombination epoch to cosmic reionization 4
   1.2 Open questions 7
      1.2.1 Primordial magnetic fields 8
      1.2.2 Dark matter decay / annihilation scenarios 13
      1.2.3 Implications for the first stars 15
   1.3 Probing the early universe 16

2 Reionization - A probe for the stellar population and the physics of the early universe[1]   19
   2.1 Introduction 19
   2.2 Reionization in the early universe 22
      2.2.1 The RECFAST code 23
      2.2.2 The generalized filtering mass 26
      2.2.3 Stellar feedback 27
      2.2.4 Models for the stellar population 28
   2.3 The effect of magnetic fields on the IGM 30
      2.3.1 Ambipolar diffusion heating 30
      2.3.2 Decaying MHD turbulence 31
      2.3.3 The evolution of the IGM 32
   2.4 Implications from the dark sector 33
      2.4.1 Dark Matter annihilation 34
      2.4.2 Dark Matter decay 35

CONTENTS

	2.4.3 Dark stars	35
2.5	Constraining the parameter space with WMAP 5	36
	2.5.1 Stellar reionization with and without primordial magnetic fields	36
	2.5.2 Dark reionization scenarios	39
2.6	Conclusions and outlook	41

3 Influence of primordial magnetic fields on 21 cm emission[1]   **45**
- 3.1 Introduction . . . 45
- 3.2 The evolution of the IGM . . . 47
  - 3.2.1 The RECFAST code . . . 47
  - 3.2.2 Heating due to primordial magnetic fields . . . 50
  - 3.2.3 The filtering mass scale and stellar feedback . . . 52
- 3.3 The 21 cm background . . . 53
  - 3.3.1 The spin temperature . . . 53
  - 3.3.2 The build-up of a Lyman-α background . . . 57
  - 3.3.3 21 cm fluctuations . . . 59
- 3.4 The evolution of linear perturbations in the dark ages . . . 60
- 3.5 The 21 cm power spectrum . . . 62
- 3.6 Discussion and outlook . . . 64

4 The chemistry of the early universe and its signatures in the CMB[1]   **67**
- 4.1 Introduction . . . 67
- 4.2 Imprints from primordial molecules on the CMB . . . 70
- 4.3 Recombination and the formation of molecules in the early universe . . . 72
  - 4.3.1 Hydrogen recombination . . . 72
  - 4.3.2 H_2 chemistry . . . 74
  - 4.3.3 Deuterium chemistry . . . 75
  - 4.3.4 HeH^+ chemistry . . . 75
- 4.4 The chemical network . . . 76
- 4.5 The numerical algorithm . . . 77
- 4.6 Results from the molecular network . . . 79
- 4.7 Effects of different species on the cosmic microwave background . . . 84
  - 4.7.1 Molecular lines . . . 84
  - 4.7.2 The negative hydrogen ion . . . 86
  - 4.7.3 The negative helium ion . . . 88
  - 4.7.4 Photodissociation of HeH^+ . . . 88
  - 4.7.5 Observational relevance and results . . . 89
- 4.8 Discussion and outlook . . . 93

CONTENTS

5 Dark stars: Implications and constraints from cosmic reionization and extra-galactic background radiation[1]   97
   5.1 Introduction 97
   5.2 The models 99
     5.2.1 Main-sequence dominated models 99
     5.2.2 Capture-dominated models 100
   5.3 Reionization constraints 101
     5.3.1 General approach 101
     5.3.2 Reionization with MS-dominated dark stars 104
     5.3.3 Reionization with CD dark stars 106
   5.4 Predictions for 21 cm observations 108
   5.5 Cosmic constraints on massive dark matter candidates 110
     5.5.1 Gamma-ray constraints 111
     5.5.2 Neutrino constraints 113
     5.5.3 Emission from dark star remnants 114
     5.5.4 Dependence of dark star models on the neutralino mass 116
   5.6 Cosmic constraints on light dark matter 116
     5.6.1 511 keV emission 117
     5.6.2 Internal Bremsstrahlung 120
     5.6.3 Emission from dark star remnants 122
   5.7 Summary and discussion 123

6 Cosmic constraints rule out s-wave annihilation of light dark matter[1]   127
   6.1 Introduction 127
   6.2 Assumptions 128
   6.3 Formalism 129
   6.4 Results 130
   6.5 Implications and discussion 131

7 Probing high-redshift quasars with ALMA[1]   135
   7.1 Introduction 135
   7.2 Chemistry in high-redshift quasars 137
   7.3 Observables in the PDR 140
   7.4 Expectations for the XDR 142
     7.4.1 Implications of X-rays for molecular clouds 142
     7.4.2 Separating the XDR from a nuclear starburst 143
     7.4.3 Model predictions for XDRs of constant size 144
     7.4.4 Model predictions for variable XDR-sizes 146
   7.5 Evidence for XDRs and the interpretation of sub-mm line observations . . . 147

		7.5.1 NGC 1068	148
		7.5.2 APM 08279	148
		7.5.3 SDSS J114816.64+525150.3	149
	7.6	The expected number of sources	151
		7.6.1 Black hole growth at high redshift	152
		7.6.2 Estimates on the number of sources	153
	7.7	Conclusions	155
8	**Discussion and outlook**		**167**
	8.1	Summary	167
		8.1.1 Reionization constraints	167
		8.1.2 21 cm observations	168
		8.1.3 Constraints from the cosmic gamma-ray background and the atmospheric neutrino background	169
		8.1.4 Signatures of primordial molecules in the CMB	170
		8.1.5 High redshift quasar observations with ALMA	170
	8.2	Open questions	172
		8.2.1 Physics of the early universe	172
		8.2.2 Reionization and the stellar population	173
		8.2.3 High-redshift quasars	174
	8.3	Outlook	175
		8.3.1 Magnetic fields	175
		8.3.2 Dark matter annihilation	176
		8.3.3 The first quasars	176
A	**Free-free transitions involving H^{-1}**		**179**
B	**Reaction rates for primordial chemistry[1]**		**183**
Bibliography			**191**

List of Figures

1.1 The formation of the first star, as calculated by Abel et al. [2002]. Credit & Copyright: Tom Abel (Stanford) & Ralf Kaehler (ZIB). 5

1.2 The observed 511 keV emission from the Galactic center: Evidence for dark matter annihilation? Credit: Ködlseder et al., A&A, 441, 513, 2003, reproduced with permission © ESO. 13

2.1 Evolution of different quantities as a function of redshift. a) The effective gas temperature. b) The effective ionized fraction. c) The filtering mass. d) The comoving magnetic field strength. 32

2.2 The reionization optical depth for Pop. III stars (model A) in the presence of primordial magnetic fields. The contour lines are equally spaced around $\tau_{re} = 0.87$ with $\Delta\tau = 0.017$, corresponding to the different σ-errors of the measurement. Magnetic fields higher than 20 nG are clearly excluded by the optical depth alone. The transition between stellar reionization and collisional ionization occurs at about 5 nG, providing a more stringent limit if metal enrichment is required. 37

2.3 The reionization optical depth for Pop. III stars (model B) in the presence of primordial magnetic fields. Magnetic fields higher than 20 nG are clearly excluded by the optical depth alone. The transition between stellar reionization and collisional ionization occurs at about 5 nG. We can further exclude magnetic fields between 2 and 5 nG within 3-σ, as too high star formation efficiencies would be required to obtained the measured optical depth. . . . 37

2.4 The reionization optical depth for a mixed population (model C) in the presence of primordial magnetic fields. Magnetic fields higher than 20 nG are clearly excluded by the optical depth alone. The transition between stellar reionization and collisional ionization occurs at about 5 nG. In addition, magnetic fields between 0.7 and 5 nG can be excluded, as they would require unreasonably high star formation efficiencies for the measured optical depth. 38

LIST OF FIGURES

2.5 The reionization optical depth for Pop. II stars (model D) in the presence of primordial magnetic fields. For magnetic fields lower than 5 nG, stellar reionization dominates, but unreasonably high star formation efficiencies would be required to reconcile the measured optical depth. For stronger magnetic fields, collisional reionization dominates, which can be excluded if metal-enrichment is imposed. Magnetic fields larger than 20 nG can be excluded from the optical depth alone. 38

2.6 The reionization optical depth for Pop. III stars (model B) in the presence of dark matter annihilation. At $\langle \sigma v \rangle / m_{DM} \sim 10^{-33}$ cm^3/s/eV, secondary ionization of the annihilation products starts to dominate over stellar reionization. Values higher than 3×10^{-33} cm^3/s/eV are ruled out by the WMAP 5-year data. 40

2.7 The reionization optical depth for Pop. III stars (model B) in the presence of dark matter decay. For lifetimes lower than 3×10^{24} s, secondary ionization of the annihilation products starts to dominate over stellar reionization. Lifetimes lower than 3×10^{23} s are ruled out by the WMAP 5-year data. ... 40

3.1 Evolution of gas temperature (upper panel) and ionized fraction (lower panel) in the medium that was not yet affected by UV feedback. The details of the models are given in Table 3.1. 48

3.2 Evolution of the filtering mass scale for different comoving magnetic field strengths. 49

3.3 CMB, gas and spin temperature in the gas unaffected by reionization, for a star formation efficiency $f_* = 10^{-3}$, in the zero field case (upper panel) and for a field strength of 0.8 nG (lower panel). In the absence of magnetic fields, gas and CMB are strongly coupled due to efficient Compton scattering of CMB photons. At about redshift 200, the gas temperature decouples and evolves adiabatically, until it is reheated due to X-ray feedback. The spin temperature follows the gas temperature until redshift ~ 100. Collisional coupling then becomes less effective, but is replaced by the Wouthuysen-Field coupling at $z \sim 25$. In the presence of magnetic fields, additional heat goes into the gas, it thus decouples earlier and its temperature rises above the CMB. Wouthuysen-Field coupling becomes effective only at $z \sim 15$, as the high magnetic Jeans mass delays the formation of luminous sources significantly. 54

3.4 The mean evolution in the overall neutral gas, calculated for the models described in Table 3.1. Top: The spin temperature. Middle: The mean brightness temperature fluctuation. Bottom: The frequency gradient of the mean brightness temperature fluctuation. 55

LIST OF FIGURES

3.5 The evolution of the 21 cm large-scale power spectrum and the growth of large-scale fluctuations in temperature and ionization for the models given in Table 3.1. Top: The evolution of the 21 cm power spectrum, normalized with respect to the baryonic power spectrum at redshift 0. Middle: The evolution of the ratio $g_T = \delta_T/\delta$. Bottom: The evolution of the ratio $g_i = \delta_{x_i}/\delta$. . 63

4.1 The idealized three-level hydrogen atom (ground state $1s$, excited states $2s$ and $2p$, continuum), and the relevant transitions. 72
4.2 Results for the evolution of H_2^+ and H^-. 79
4.3 Results for the evolution of D^+ and HeH^+. 80
4.4 Results for the evolution of HD^+ and D^-. 80
4.5 Results for those species which freeze-out and are not in chemical equilibrium. 81
4.6 HeH^+ photodissociation cross section obtained from detailed balance. . . . 88
4.7 The HeH^+ optical depth, both corrected and uncorrected for stimulated emission. The contributions from pure rotational and ro-vibrational transitions are given separately. 89
4.8 The HD^+ optical depth, both corrected and uncorrected for stimulated emission. The contributions from pure rotational and ro-vibrational transitions are given separately. 90
4.9 The absorption optical depth due to different processes: the free-free processes of H^- and He^-, the bound-free process of H^-, and the photodissociation of HeH^+. Clearly, the total optical depth is dominated by the processes involving H^-. 91
4.10 The relative change in the CMB temperature due to the presence of H^-. We further plot the optical depths due to absorption, spontaneous and stimulated emission for the bound-free process, as their overlap explains some features in the temperature change. 92

5.1 The evolution of the effective ionized fraction x_{eff}, for reionization models with main-sequence dominated dark stars (see Table 5.1). Models MS 1 and MS 2 can be ruled out by reionization constraints, while models MS 3 and MS 4 require a sudden increase in the star formation rate by a factor of 30 at redshift 6.5. It appears more realistic to assume lower masses and star formation efficiencies to reconcile dark star models with observations. . . . 103
5.2 The evolution of the effective ionized fraction x_{eff}, for reionization models with capture-dominated dark stars (see Table 5.2). Models CD 1a, CD 1b, CD 2a, CD 2b and CD 3 are ruled out due to reionization constraints, while the remaining models require an artificial star burst. 106

LIST OF FIGURES

5.3 21 cm signatures of double-reionization scenarios (here MS 4 from Table 5.1). Given is the evolution after the first reionization phase, when the H gas is heated from the previous ionization. Top: HI gas temperature, here identical to the spin temperature. Middle: Expected mean 21 cm brightness fluctuation. Bottom: Frequency gradient of the mean 21 cm brightness fluctuation. ... 109

5.4 The predicted gamma-ray background due to direct annihilation into gamma-rays in the presence of adiabatic contraction during the formation of dark stars, and the background measured by EGRET (squares) [Strong et al., 2004]. One finds two peaks in the annihilation background for a given particle mass: One corresponding to annihilation at redshift zero, and one corresponding to the redshift where the enhancement from adiabatic contraction was strongest. ... 111

5.5 The predicted neutrino background due to direct annihilation into neutrinos in the presence of adiabatic contraction during the formation of dark stars, and the atmospheric neutrino background [Honda et al., 2004]. One finds two peaks in the annihilation background for a given particle mass: One corresponding to annihilation at redshift zero, and one corresponding to the redshift where the enhancement from adiabatic contraction was strongest. ... 113

5.6 The *maximum* gamma-ray background due to direct annihilation into gamma-rays in the remnants of dark stars, and the background measured by EGRET [Strong et al., 2004]. The actual contribution to the gamma-ray background is highly model-dependent (see discussion in the text). ... 114

5.7 The predicted X-ray background due to 511 keV emission for different dark matter particle masses. Solid lines: Enhanced signal due from adiabatic contraction, dotted lines: Conventional NFW profiles. The observed X-ray background from the HEAO experiments (squares) [Gruber et al., 1999] and Swift/BATSE (triangles) [Ajello et al., 2008] is shown as well. The comparison yields a lower limit of 10 MeV on the dark matter mass for the adiabatically contracted profiles, and 7 MeV for standard NFW halo profiles. ... 118

5.8 The predicted gamma-ray background due to bremsstrahlung emission for different dark matter particle masses. Solid lines: Enhanced signal due from adiabatic contraction, dotted lines: Conventional NFW profiles. The lines overlap almost identically, as the main contribution comes from redshift zero, where the clumping factor is large and dark stars are assumed not to form. The observed gamma-ray background from the HEAO experiments (squares) [Gruber et al., 1999], Swift/BATSE (triangles) [Ajello et al., 2008], COMPTEL (crosses) [Kappadath et al., 1996] and SMM (plusses) [Watanabe et al., 1999] is shown as well. ... 121

LIST OF FIGURES

5.9 The upper limit of X-ray radiation due to dark star remnants. The observed X-ray background from the HEAO experiments (squares) [Gruber et al., 1999] and Swift/BATSE (triangles) [Ajello et al., 2008] is shown as well. Only for very low dark matter particle masses, the upper limit is somewhat higher than the observed background. However, the actual contribution may be lower by some orders of magnitude (see discussion in the text). 122

6.1 The predicted gamma-ray background due to internal bremsstrahlung emission for different dark matter particle masses. In every case, we adopted a clumping factor that yields the maximum allowed background. We compare with the observed gamma-ray background from COMPTEL (crosses) [Kappadath et al., 1996], SMM (plusses) [Watanabe et al., 1999] and EGRET (squares) [Strong et al., 2004]. 131

6.2 The constraint on the clumping factor C_0 at redshift zero due to internal bremsstrahlung emission. For comparison, we show the weaker constraint due to 511 keV emission. The forbidden region is shaded. We also show the upper mass limit from Galactic center observations calculated by Beacom & Yüksel [2006] as well as the minimal clumping factor $C_0 = 10^5$. The combination of these constraints shows that light dark matter models with s-wave annihilation are ruled out. 132

7.1 The heating rates per hydrogen atom due to X-ray absorption (XDR contribution), absorption of soft UV-photons (PDR contribution) and total heating rate as a function of radius. The calculation assumes a 10^7 $M_\odot$ black hole, with 3% of its Eddington luminosity being emitted in a hard spectral component between 1 and 100 keV with a spectral slope of -1. The strength of the soft UV radiation field is taken as $G_0 = 10$ in Habing units and the effective density $n_{\rm eff} = 10^5$ cm^{-3}. For such a configuration, the XDR contribution clearly dominates within the central 100 pc. 137

7.2 The expected size of the X-ray dominated region in pc, for a black hole with 10^7 $M_\odot$ with a spectral slope of -1, as a function of the soft UV radiation field G_0 in Habing units and the effective number density $n_{\rm eff}$. The straight lines assume the power-law between 1 and 5 keV, while the dotted lines assume it between 1 and 100 keV. 138

7.3 A model for the X-ray chemistry in NGC 1068. The adopted flux impinging on the cloud is 170 erg s^{-1} cm^{-2}. The adopted density is 10^5 cm^{-3}. Top: The abundances of different species as a function of column density. Middle: The low-J CO lines as a function of column density. Bottom: The high-J CO lines as a function of column density. 157

LIST OF FIGURES

7.4 The X-ray chemistry in a system with X-ray flux of 1 erg s^{-1} cm^{-2} impinging on the cloud. The adopted density is 10^5 cm^{-3}. Top: The abundances of different species as a function of column density. Middle: The low-J CO lines as a function of column density. Bottom: The high-J CO lines as a function of column density. 158

7.5 The X-ray chemistry in a system with X-ray flux of 10 erg s^{-1} cm^{-2} impinging on the cloud. The adopted density is 10^5 cm^{-3}. Top: The abundances of different species as a function of column density. Middle: The low-J CO lines as a function of column density. Bottom: The high-J CO lines as a function of column density. 159

7.6 A comparison of the CO line SED in case of an intense starburst with $G_0 = 10^5$ with the corresponding SED for molecular clouds under X-ray irradiation, for different X-ray fluxes in erg s^{-1} cm^{-2}. The spectra are normalized such that they have the same intensity in the 10th transition. If the impinging X-ray flux is at least 2.8 erg s^{-1} cm^{-2}, observations of the 15th CO rotational transition can clearly discriminate between PDR and XDR chemistry. 160

7.7 The expected flux in mJy for high-J CO lines, for a central XDR of 200 pc, and molecular clouds of 10^5 cm^{-3}, as a function of X-ray luminosity and cloud column density. We focus on lines that fall in ALMA band 6, which offers a good compromise between angular resolution and sensitivity. For a source at $z = 5$ (solid line), this corresponds to the (10-9) CO transition, for a source at $z = 8$, it corresponds to the (14-13) CO transition. 160

7.8 The expected flux in mJy for the [CII] 158 μm line, for a central XDR of 200 pc, and molecular clouds of 10^5 cm^{-3}, as a function of X-ray luminosity and cloud column density. The solid line corresponds to a source at $z = 5$, and the dashed line to a source at $z = 8$. At $z = 5$, the line is redshifted into ALMA band 7, with a sensitivity of 0.08 mJy (1 day, 3σ, 300 km/s). At $z = 8$, it falls into ALMA band 6 with a sensitivity of 0.04 mJy. 161

7.9 The expected flux in mJy for the [CII] 158 μm line, for a central XDR of 200 pc, and molecular clouds of 10^4 cm^{-3}, as a function of X-ray luminosity and cloud column density. The solid line corresponds to a source at $z = 5$, and the dashed line to a source at $z = 8$. At $z = 5$, the line is redshifted into ALMA band 7, with a sensitivity of 0.08 mJy (1 day, 3σ, 300 km/s). At $z = 8$, it falls into ALMA band 6 with a sensitivity of 0.04 mJy. 161

LIST OF FIGURES

7.10 The expected flux in mJy for the [OI] 63 μm line, for a central XDR of 200 pc, and molecular clouds of 10^5 cm^{-3}, as a function of X-ray luminosity and cloud column density. The solid line corresponds to a source at $z = 5$, and the dashed line to a source at $z = 8$. At $z = 5$, the line is redshifted into ALMA band 10, with a sensitivity of 0.2 mJy (1 day, 3σ, 300 km/s). At $z = 8$, it falls into ALMA band 8 with a sensitivity of 0.13 mJy. 162

7.11 The expected flux in mJy for the [OI] 63 μm line, for a central XDR of 200 pc, and molecular clouds of 10^4 cm^{-3}, as a function of X-ray luminosity and cloud column density. The solid line corresponds to a source at $z = 5$, and the dashed line to a source at $z = 8$. At $z = 5$, the line is redshifted into ALMA band 10, with a sensitivity of 0.2 mJy (1 day, 3σ, 300 km/s). At $z = 8$, it falls into ALMA band 8 with a sensitivity of 0.13 mJy. 162

7.12 The expected flux in mJy for the [OI] 146 μm line, for a central XDR of 200 pc, and molecular clouds of 10^5 cm^{-3}, as a function of X-ray luminosity and cloud column density. The solid line corresponds to a source at $z = 5$, and the dashed line to a source at $z = 8$. At $z = 5$, the line is redshifted into ALMA band 7, with a sensitivity of 0.37 mJy (1 day, 3σ, 300 km/s). At $z = 8$, it falls into ALMA band 6 with a sensitivity of 0.225 mJy. 163

7.13 The expected flux in mJy for the CO (14-13) transition (solid line), [CII] 158 μm (dotted line), [OI] 63 μm (dashed line) and [OI] 146 μm (dot-dashed line) emission for a quasar at $z = 5$, as a function of X-ray luminosity and average density. We assume a typical cloud column density of 10^{23} cm^{-2}, and a soft-UV field $G_0 = 100$. 163

7.14 The expected flux in mJy for the CO (14-13) transition (solid line), [CII] 158 μm (dotted line), [OI] 63 μm (dashed line) and [OI] 146 μm (dot-dashed line) emission for a quasar at $z = 8$, as a function of X-ray luminosity and cloud column density. We assume a typical cloud density of 10^4 cm^{-3}, and a soft-UV field $G_0 = 100$. 164

7.15 A sketch for a situation with an inhomogeneous XDR, motivated by the observations of Galliano et al. [2003] in NGC 1068. While X-rays are shielded by a central absorber along the line of sight, they may stimulate emission in molecular clouds in the perpendicular direction with the typical characteristics of XDRs. 164

7.16 The average accretion history of a black hole with 10^7 $M_\odot$ at $z = 5$, depending on a coefficient p which is a function of the conversion of rest-mass energy to luminous energy, the Eddington ratio and the duty cycle. 165

xv

LIST OF FIGURES

7.17 Estimates for the number of sources in a redshift interval of $\Delta z = 0.5$ within a solid angle of $(1')^2$. We show extrapolations of the results by Shankar et al. [2009] and Treister et al. [2009] to high redshift, for luminosities larger than 10^{44} erg s^{-1}. In addition, we give an estimate based on the number of high-redshift black holes required in order to produce the present-day $10^9\ M_\odot$ black holes. As discussed in the text, the actual number of black holes should lie in between these estimates. 165

A.1 The free-free absorption coefficient of H$^-$ for 10^2, 10^3, 10^4 and 10^5 GHz, as a function of temperature. Given are the fits of John [1988] and Gingerich [1961] for the high-temperature regime, the calculation of Dalgarno & Lane [1966] based on effective range theory as well as the new calculation of this work for the low-temperature regime up to 2000 K. 180

List of Tables

2.1 Summary of adopted stellar models. The first column gives the model name. The second column indicates the stellar populations that contribute to reionization. The third column gives the number of ionizing photons per baryon used in Eq (2.16). The fourth column lists the adopted escape fractions. Model A is a highly extreme case in which we assume all ionizing photons come from massive Pop III stars and can escape the star-forming halo. In model B photons can escape efficiently only from halos less massive than a corresponding virial temperature of 10^4 K, while for higher-mass halos, this fraction is reduced to 10%. In model C we assume that Pop III as well as Pop II stars contribute to reionization, and we adopt the same escape probabilities as in B. Model D is another extreme case which assumes that all ionizing photons come from low-mass Pop II stars, with escape fraction 6%. . . . 29

3.1 A list of models for different co-moving magnetic fields and star formation efficiencies, which are used in several figures for illustrational purposes. We give the comoving magnetic field B_0 and the star formation efficiency f_*. For illustration purposes, all models assume a population of massive Pop. III stars. The amount of Lyman α photons produced per stellar baryon would be larger by roughly a factor of 2 if we were to assume Pop. II stars. Assuming that the same amount of mass goes into stars, the coupling via the Wouthuysen-Field effect would start slightly earlier, but the delay due to magnetic fields is still more significant. 59

4.1 Freeze-out values of free electrons, H_2 and HD at z=100. 81
4.2 Freeze-out values of free electrons, H_2 and HD at z=10. 81
4.3 Abundances of D^+ and HeH^+ at z=100. 82
4.4 Abundances of D^+ and HeH^+ at z=10. 82

LIST OF TABLES

5.1 Reionization models for MS-dominated dark stars. The parameters $z_{\text{Pop II}}$ and z_{burst} give the transition redshifts to a mode of Pop. II star formation and to the sudden star burst, while τ_{reion} is the calculated reionization optical depth and z_f the redshift of full ionization. 105

5.2 Reionization models for CD dark stars stars. The number of ionizing photons was determined from the work of Yoon et al. [2008]. The parameters $z_{\text{Pop II}}$ and z_{burst} give the transition redshifts to a mode of Pop. II star formation and to the sudden star burst, while τ_{reion} is the calculated reionization optical depth and z_f the redshift of full ionization. The calculation assumes a spin-dependent scattering cross section of 5×10^{-39} cm^2. As stellar models depend on the product of this cross section with the threshold dark matter density, the effect of a lower scattering cross section is equivalent to a smaller threshold density. 107

7.1 The main observables for ALMA in PDRs at high redshift. This specific example assumes a galaxy with a starburst as in NGC 1068 placed at $z = 8$. We also give the minimal redshift from which the lines would be redshifted into the ALMA bands. 141

7.2 Frequency range, angular resolution θ_{res} at the largest baseline, line sensitivity S_l for a linewidth of 300 km/s and continuum sensitivity S_c for 3σ detection in one hour of integration time and primary beam size θ_{beam}. 3 more bands might be added in the future, band 1 around 40 GHz, band 2 around 80 GHz and band 10 around 920 GHz, which will have similar properties as the neighbouring bands. 141

B.1 Collisional and radiative rates. 184
B.1 Collisional and radiative rates in cgs units. 185
B.1 Collisional and radiative rates in cgs units. 186
B.1 Collisional and radiative rates in cgs units. 187

1

Introduction

While a standard model of cosmology has been established, several open questions remain regarding the universe in particular at early times. Such questions concern, for instance, the presence of primordial magnetic fields as well as the true nature of dark matter, but also the origin of the observed supermassive black holes at $z \sim 6$ [Fan et al., 2001] and their high metallicities [Pentericci et al., 2002]. Observational progress makes it possible to constrain some of the physics of the early universe in more detail, like the strength of primordial magnetic fields and the effects of dark matter annihilation / decay. I will further demonstrate that future telescopes like ALMA and JWST are capable of observing active galaxies far beyond redshift 6, as long as their black hole has at least 10^6 $M_\odot$. In about five years time, it is therefore possible to detect the progenitors of supermassive black holes at high redshift and to follow the cosmic evolution of the black hole population.

I will first review the thermal history of the universe in Chapter 1.1, and discuss some of the open questions and their consequences in Chapter 1.2. A brief overview how these questions can be probed in more detail is given in Chapter 1.3, with a more detailed description presented in the following chapters.

1.1 The timeline of cosmology

During the last decades, a standard model of cosmology has been established, assuming that the universe is homogeneous and isotropic on scales larger than 100 Mpc, such that it can be described by the Friedmann-Lemaître-Robertson-Walker metric. It originates from an early hot phase in which radiation dominates and the light elements are formed. The composition of the universe has been measured with high accuracy [Komatsu et al., 2008]. At $z = 0$, only $\sim 5\%$ of the total mass-energy density is due to baryons, while 23% are due to a pressureless component of dark matter and 72% are due to a component with negative

1. INTRODUCTION

pressure that is usually referred to as dark energy, currently consistent with a cosmological constant. Structure formation models require dark matter to be cold, i. e. non-relativistic.

This scenario in which the universe consists of baryons, cold dark matter (CDM) and a cosmological constant (Λ) is usually termed the ΛCDM model. This is the current standard model of cosmology. While we know the composition of the universe, the nature of dark matter and dark energy is however not yet understood. In this section, I will describe the timeline of cosmology as expected in the standard model, as well as some of the remaining uncertainties, in particular with respect to the nature of dark matter and the putative presence of primordial magnetic fields. Of course, other uncertainties are present as well, for instance regarding the nature of dark energy. The latter is however not explored further in this work. The history of the universe and its implications for the generation of magnetic fields is discussed in the following sections.

1.1.1 From the Big Bang to cosmic recombination

As recently confirmed by the new WMAP 5 year data, the universe today appears close to being spatially flat, and observations of the cosmic microwave background show that the universe has been highly homogeneous at the epoch of recombination, over scales considerably larger than the cosmological horizon [Komatsu et al., 2008]. These observational facts, commonly referred to as the flatness and the horizon problem, appeared hard to explain in the cosmological framework, unless an early inflationary epoch is introduced. In this epoch, the universe is in a phase of rapid and accelerated expansion, with its size increasing by a factor of $\sim 10^{30}$. Most inflationary models assume that a scalar field, the inflaton field, dominates the energy budget of the universe during this epoch. It evolves in a rather flat potential, such that its kinetic energy is considerably smaller than the potential energy, giving rise to an equation of state with negative pressure and the accelerated expansion of the universe.

This scenario was first introduced by Guth [1981]. It was then realized that inflation also provides a way to generate large-scale structures in the universe from small quantum fluctuations [Hawking, 1982; Starobinskij, 1982; Guth & Pi, 1982]. A review about more recent developments is given by Kinney [2003]. Inflationary scenarios generally predict a primordial density power spectrum which is almost scale-invariant, which is in agreement with current data. Since the epoch of inflation must eventually come to an end and the universe needs to be re-heated after this era of rapid expansion, it is generally assumed that the inflaton field eventually evolves toward a minimum in its potential where it oscillates and decays due to some couplings to ordinary matter, thus increasing the entropy and heating the universe to extremely highly temperatures.

After reheating, the universe continues to expand and thus cools adiabatically. At temperatures of $\sim 10^{12} - 10^{16}$ TeV, it is expected to be in the grand unification epoch, in which

1.1 The timeline of cosmology

the weak, the strong and the electromagnetic interaction are unified [Roos, 2003]. Such scenarios are suggested for instance in supersymmetric extensions of the standard model [Drees, 1996]. The physics of this epoch is still highly uncertain and subject to speculative suggestions in particle physics. It is generally assumed that this phase is also responsible for the observed assymetry between matter and anti-matter. In order to produce such an asymmetry, a grand unification theory generally needs to fulfill three conditions:

1. It must contain reactions that violate baryon number conservation.

2. The theory must not be symmetric with respect to charge conjugation as well as the combination charge conjugation and parity change.

3. The relevant processes must occur out of thermal equilibrium.

While the first condition is obvious, the second must be required as well. Otherwise, each baryon-violating reaction would be matched by equally frequent reactions that violate baryon number in the other direction, such that the net effect is zero. Finally, a deviation from thermal equilibrium must be required, otherwise there would be no net effect because of the inverse reactions. Since most of the matter and anti-matter particles will annihilate during the further evolution of the universe, the generation of a small asymmetry is sufficient to strongly suppress the abundance of anti-matter at late times.

Once the universe cools to ~ 100 GeV, it is in the energy range of the standard model of particle physics [Roos, 2003]. At this scale, the electroweak symmetry is broken, and the W and Z vector bosons that mediate the weak interaction become massive particles. Such symmetry breaking may occur through a phase transition of first or second order. During a first order phase transition, the relevant field, here the Higgs field, tunnels through a potential barrier, while evolving smoothly from one state to another during a second-order phase transition. Lattice computations in the framework of the standard model of particle physics give strong evidence against a first-order phase transition [Kajantie et al., 1996], though it remains a viable possibility if supersymmetric extensions of the standard model are considered [e. g. Dine et al., 1992; Espinosa et al., 1993]. The electroweak baryogenesis scenario requires a first-order phase transition [Riotto & Trodden, 1999].

At an energy scale of ~ 150 MeV, the phase transition between low-energy hadronic physics and QCD occurs [Kajantie & Kurki-Suonio, 1986]. At higher energies, quarks are asymptotically free and form what is called quark matter, and at this temperature they are bound into nucleons [Bonometto & Pantano, 1993]. Typically, a first-order phase transition takes place by bubble nucleation.

At even lower temperatures and densities, neutrinos will decouple from the plasma. This occurs at a temperature of 3.5 MeV for μ and τ neutrinos, and at 2.3 MeV for electron-neutrinos. At energies of ~ 0.1 MeV, the universe is in the epoch of Big Bang nucleosynthesis (BBN). In this epoch, the light elements are formed within approximately three minutes.

1. INTRODUCTION

Big Bang nucleosynthesis was first suggested by Gamow [1946], and developed further in various works [e. g. Alpher et al., 1948; Hayashi, 1950]. The theory predicts mass abundances of ~ 75% for hydrogen, 25% for helium, 0.01% for deuterium and trace amounts of lithium and beryllium. The generally good agreement of observed and predicted abundances is one of the outstanding successes of the hot Big Bang theory. The observed lithium abundance does not match these predictions very well, which may be due to further evolution during stellar nucleosynthesis. A recent comparison of observed and predicted abundances is given by Miele & Pisanti [2008].

As the universe cools further, it eventually reaches temperatures where the electrons and ions combine. This is referred to as the epoch of recombination, which was predicted by Peebles [1968] and Zeldovich et al. [1969]. While one would naively expect the recombination of hydrogen at temperatures of ~ 10^4 K, the process is delayed until temperatures of ~ 3000 K due to the large photon-to-baryon ratio, which leads to a sufficient number of ionizing photons from the high-energy tail to keep the universe ionized until $z \sim 1100$. At recombination, the universe becomes transparent to the photons that form the cosmic microwave background (CMB). The detection of this background by Penzias & Wilson [1965] and its correct interpretation by Dicke et al. [1965] was a further milestone of cosmology. Its spectrum appears as a perfect blackbody at a temperature of 2.725 K [Fixsen & Mather, 2002]. Its power spectrum is now an important tool to derive for instance the cosmological parameters, the normalization of the dark matter power spectrum and the reionization optical depth. Ultimately, the CMB is expected to have deviations from a blackbody spectrum, for instance due to the 21 cm line of atomic hydrogen or various molecular line. This possibility is explored in more detail in Chapter 3 and 4.

1.1.2 From the postrecombination epoch to cosmic reionization

The universe after recombination is almost neutral, largely homogeneous and of primordial composition. Initially, the gas temperature is still coupled to the CMB temperature by Compton scattering. This coupling becomes inefficient at $z \sim 200$, and the gas temperature then evolves adiabatically and drops below the CMB temperature. With decreasing density, the recombination timescale becomes significantly larger than the expansion timescale, such that the universe does not become fully neutral, but rather retains a relic electron abundance of ~ 2×10^{-4} [Peebles, 1968; Zeldovich et al., 1969; Seager et al., 2000].

As recognized initially by Saslaw & Zipoy [1967], such an electron abundance leads to the formation of molecular hydrogen in the intergalactic medium (IGM) and the first collapsing objects. In the IGM, H_2 formation starts at $z \sim 300$ via the H_2^+ formation channel and the abundance increases at $z \sim 100$ via the H^- formation channel [Puy et al., 1993; Stancil et al., 1998; Galli & Palla, 1998]. In the first collapsing gas clouds at $z \sim 25$, the H_2 abundance increases further due to the higher densities. In the absence of metal coolants, this

1.1 The timeline of cosmology

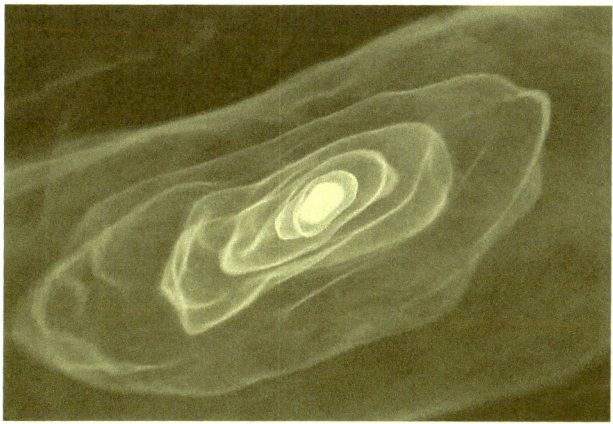

Figure 1.1: The formation of the first star, as calculated by Abel et al. [2002]. Credit & Copyright: Tom Abel (Stanford) & Ralf Kaehler (ZIB).

is crucial in particular for the first minihalos with virial temperatures below 10^4 K [Tegmark et al., 1997], since they cannot cool by atomic hydrogen.

A formalism to estimate the number density of halos as a function of their mass at a given redshift has been developed by Press & Schechter [1974] and was improved by Sheth et al. [2001]. For the current cosmological parameters, this framework predicts the first starforming halos around $z \sim 25$. Numerical simulations indicate that gas collapse in these minihalos leads to the formation of stars with $\sim 100\ M_\odot$ [Abel et al., 2002; Bromm & Larson, 2004; Yoshida et al., 2006]. A result from such high-resolution numerical simulations is shown in Fig. 1.1. Such stars provide strong sources of ionizing radiation [Bromm et al., 2001; Schaerer, 2002], and are required to explain the observed reionization optical depth [Wyithe & Loeb, 2003; Schleicher et al., 2008b]. Stars forming in primordial gas are generally referred to as Population III stars (Pop. III), specifically Pop. III.1 if they form in neutral gas and Pop. III.2 if they form in previously ionized gas (see below). This terminology originates from the classification of stars in the Milky Way, where metal-rich stars like the sun are called Pop. I stars and metal-poor stars are termed Pop. II stars.

The evolution of the subsequent stellar population is more difficult to infer, since the first stars will affect their environment by mechanical, chemical and radiative feedback. In the range of $140 - 260\ M_\odot$, it is expected that they explode as a pair-instability supernova, while one expects a conventional core-collapse supernova for masses smaller than $40\ M_\odot$. Inbetween these two ranges, the star would collapse to form a black hole [Heger & Woosley,

1. INTRODUCTION

2002]. Mechanical feedback from violent supernova explosions changes the gas distribution and thus the initial conditions for subsequent generations of stars [Greif et al., 2007]. It also affects the ionization degree and leads to metal enrichment. While a detailed calculation of the gas fallback after a supernova explosion and the subsequent initial conditions for star formation is still outstanding, it has been demonstrated that the correct initial conditions are indeed crucial for the further evolution of the gas [Jappsen et al., 2008]. Once a critical metallicity is reached, arguable at $\sim 10^{-5}Z_\odot$ or $10^{-3}Z_\odot$, the gas fragments and star formation occurs according to a standard Pop. II initial mass function (IMF) [Bromm et al., 2001; Schneider et al., 2003; Clark et al., 2008].

Radiative feedback will affect the next generation of stars in various ways. Ionization of the gas by UV photons enhances the ability of the gas to form molecular hydrogen and HD, thus providing an efficient cooling mechanism down to the CMB floor. This leads to smaller Jeans masses and typically decreases the stellar mass by an order of magnitude [Yoshida et al., 2007a,b]. These are the so-called Pop. III.2 stars. Softer UV photons in the Lyman-Werner (LW) range of 11.2 – 13.6 eV can dissociate molecular hydrogen by the Solomon process [Stecher & Williams, 1967]. This effect was explored a couple of years ago by Haiman et al. [2000], Glover & Brand [2001] and Machacek et al. [2001], indicating that it may provide an efficient mechanism to suppress subsequent star formation by destroying the main coolant. However, more recent work shows that the build-up of such a LW-background is a self-regulated process, and the IGM may soon be optically thick to such radiation while H_2 builds up in relic HII regions [Johnson et al., 2007a, 2008]. It is thus conceivable that the latter is only a minor correction to the total star formation rate.

In addition, the universe could be magnetized by magnetic fields generated in winds and supernova explosions from the first stars [Rees, 1987; Kandus et al., 2004]. To my knowledge, the effect of such feedback on the subsequent generations of stars has not yet been explored. For the very first stars, people started to consider the consequences of magnetic fields generated during protostellar collapse (see discussion in Chapter 1.2.3 and Chapter 8.2.2), but the results are still uncertain and controversial.

Toward $z \sim 10$, halos grow in size and may harbor more massive galaxies. The so-called atomic-cooling halos, those with virial temperatures above 10^4 K can cool via atomic hydrogen and so do not strictly require the presence of molecular hydrogen to collapse, though simulations indicate that this coolant is also important in these systems [Wise & Abel, 2007a; Greif et al., 2008]. They are much more turbulent than the first minihalos and consist of a two-phase medium of both hot and cold gas [Greif et al., 2008]. The expected stellar masses in such galaxies are currently unclear, but may strongly depend on the metallicity [Clark et al., 2008]. It has also been suggested that these halos could provide the seeds for the first supermassive black holes [Eisenstein & Loeb, 1995; Koushiappas et al., 2004; Begelman et al., 2006; Spaans & Silk, 2006; Dijkstra et al., 2008]. However, important uncertainties remain regarding the formation process. For instance, Lodato & Natarajan [2006] argue that the gas in such halos should fragment if H_2 cooling is efficient, and Omukai et al. [2008] and Jappsen et al. [2008] suggest that a non-zero metallicity will lead to fragmentation.

The first stars and the first galaxies constitute the sources of light that ultimately lead to the reionization of the universe. Simulations of stellar reionization further show that the process is highly inhomogeneous and based on the growth and merging of ionized bubbles around the first stellar sources [Gnedin, 2000; Ciardi et al., 2003; Kohler et al., 2005], and semi-analytic models have been introduced which provide an effective prescription of such processes [Shapiro & Giroux, 1987; Haiman & Loeb, 1997a; Barkana & Loeb, 2001; Loeb & Barkana, 2001; Choudhury & Ferrara, 2005; Schleicher et al., 2008b; Schneider et al., 2008]. X-ray feedback from high-redshift quasars may only constitute a minor contribution, due to the constraints from the soft X-ray background [Dijkstra et al., 2004; Salvaterra et al., 2005].

At present, reionization is constrained to end by $z \sim 6$ from the spectra of high-redshift quasars [Becker et al., 2001], which show strong Lyman α absorption features at higher redshift, as well as the reionization optical depth due to Thomson scattering on free electrons. This boosts the polarization of the CMB on large scales, while it suppresses polarization on small scales [Zaldarriaga, 1997]. The reionization optical depth could be derived for the first time from the WMAP 1-year data [Kogut, 2003], and has now been calculated with much higher accuracy from the five-year data to be $\tau_e = 0.087 \pm 0.017$.

Black holes with $\sim 10^9 \, M_\odot$ have already been detected at $z \sim 6$, and their accretion disks were found to have super-solar metallicities even at such redshifts [Pentericci et al., 2002; Juarez et al., 2009]. Observationos of their spectra provide evidence for the completion of reionization near $z = 6$ [Becker et al., 2001]. To explain the origin of supermassive black holes is an important goal of cosmology and structure formation. In Chapter 7, I discuss how observations with ALMA and JWST can detect their progenitors at higher redshift.

1.2 Open questions

In this section, I discuss some of the open questions with respect to the cosmological model described above. This concerns the potential presence of primordial magnetic fields, as well as the true nature of dark matter, which may reveal itself via dark matter annihilation or decay. I review both the observational and the theoretical motivations for these scenarios. Then I discuss their impact on the formation of the first stars, before exploring the consequences in more detail in the following chapters.

1. INTRODUCTION

1.2.1 Primordial magnetic fields

To explore the potential relevance of primordial magnetic fields in more detail, I will discuss whether currently observed magnetic fields can be explained by astrophysical mechanisms in Chapter 1.2.1.1, and present a description of the mechanisms available in the era before cosmic recombination to generate primordial magnetic fields in Chapter 1.2.1.2. Some of these topics are discussed in more detail by Grasso & Rubinstein [2001].

1.2.1.1 Primordial or astrophysical?

Magnetic fields can be observed by Zeeman splitting of spectral lines, Faraday rotation measurements of polarized electromagnetic radiation passing through an ionized medium, and from synchrotron emission of free relativistic electrons. Comprehensive reviews of these topics are available from Kronberg [1994] and Zweibel & Heiles [1997]. While Zeeman splitting provides a direct way of detection, the effect is very small and only useful within our galaxy. On the other hand, synchrotron emission and Faraday rotation measurements can trace magnetic fields at large distances, but require an independent determination of the local electron density n_e. This is sometimes possible, for instance by studying X-ray emission in the very hot gas of galaxy clusters.

In our Galaxy, the average field strength is 3 – 4 μG, corresponding to approximate equipartition between the magnetic field, the cosmic rays confined in the galaxy, and the small-scale turbulent motion [Kronberg, 1994]. The field is coherent over scales of a few kpc, comparable to the galactic size, with alternating directions in the arm and interarm regions [Kronberg, 1994; Han, 2008]. For a long time, the preferred mechanism to explain these observations was the dynamo mechanism [Zeldovich et al., 1979; Vainshtein et al., 1980]. For the generation of large coherence lengths, the mean-field dynamo is particularly popular, which is derived by defining a spatial average over the induction equation on scales larger than the turbulent eddy scales [Krause & Rädler, 1980]. However, while the mean-field dynamo would operatore on large scales, small-scale fields could be produced on a much faster rate, leading to saturation before a significant large-scale field is built-up [Kulsrud et al., 1997]. Also, most dynamo models predict even parity with no reversals, contrary to observations in our Galaxy, whereas such a situation would be expected in the case of primordial fields [Grasso & Rubinstein, 2001]. Zweibel [2006] confirms that the problem of creating a large-scale field from astrophysical sources is still unresolved. Kulsrud & Zweibel [2008] consider some commonly raised arguments against primordial magnetic fields, finding that these arguments are subject to considerable uncertainties and cannot rule out the primordial case.

1.2 Open questions

Strong magnetic fields have also been detected in the inter-cluster medium (ICM) of galaxy clusters [Kim et al., 1991]. Their field strength is typically described by the phenomenological equation

$$B_{\rm ICM} \sim 1.4 \mu G \left(\frac{L}{10 \text{ kpc}} \right)^{-1/2} (h_{70})^{-1}, \tag{1.1}$$

where L is the reversal field length and h_{70} the Hubble constant normalized to 70 km s^{-1} Mpc^{-1}. Typical values of L range between 10 and 100 kpc. In the Coma cluster, a core magnetic field of ~ 8 μG tangled at scales of about 1 kpc has been detected [Feretti et al., 1995]. While it was recently shown that magnetic fields from outflows can lead to similar field strength in galaxy clusters [Donnert et al., 2009], it is not yet clear if that also explains the coherence on large scales. It appears that the results are hard to distinguish from a configuration created by primordial magnetic fields, as calculated for instance by [Dolag et al., 1999]. As shown by Banerjee & Jedamzik [2003], primordial magnetic fields generated during cosmological phase transitions could both explain the observed field strength and the large-scale coherence.

Magnetic fields have also been detected in very distant objects; for instance radio quasars near $z \sim 2$ appear to have field strength of a few μG [Kronberg, 1994]. That by itself does not necessarily speak for or against primordial magnetic fields, as their coherence lengths is poorly known and quasars generally appear as highly evolved also in metallicity [Juarez et al., 2009]. However, Bernet et al. [2008] recently showed that quasars with strong Mg$^+$ absorption lines are unambiguously associated with large Faraday rotation measurements. As Mg$^+$ absorption occurs in the halos of normal galaxies along the line of sight of quasars, they found evidence for magnetic field strengths of a few μG in normal galaxies at $z \sim 3$, with considerably less time available for a dynamo mechanism to operate. Indeed, as noted already by Grasso & Rubinstein [2001], such field strengths can be explained by primordial fields of ~ 0.1 nG without requiring a dynamo to operate.

At even higher redshift, one can only give upper limits on the expected field strength. Measurements of the small-scale CMB anisotropy yield an upper limit of 4.7 nG to the comoving field strength on a scale of 1 Mpc [Yamazaki et al., 2006]. Constraints on the primordial field strength are also available from the measurement of σ_8, which describes the root-mean-square of matter density fluctuations in a comoving sphere of $8h^{-1}$ Mpc. However, they depend sensitively on the assumed power spectrum for the magnetic field [Yamazaki et al., 2008]. Other constraints are available from Big Bang nucleosynthesis (BBN). For instance, strong magnetic fields above 4.4×10^{13} G at the epoch of BBN significantly increase the β decay rate of neutrons, which would suppress the ^{4}He abundance with respect to the standard case [Matese & O'Connell, 1969; O'Connell & Matese, 1969]. Strong uniform fields would further affect the energy density of the electron gas due to the increase of its

1. INTRODUCTION

phase space [Matese & O'Connell, 1970]. More relevant is however the contribution of the magnetic field energy density to the total energy density, and its impact on the expansion of the universe [Greenstein, 1969].

Recent works confirm that the allowed magnetic field strength at the end of nucleosynthesis ($T = 0.01$ MeV) is $\sim 2 \times 10^9$ G, corresponding to $\sim 10^{-6}$ G today.

It is important to note that the constraints described here are rather weak, and magnetic fields of such strength can have a considerable impact on the evolution of the early universe and the formation of the first stars and galaxies. This will be described in more detail in Chapter 2 and 3, where I discuss how such constraints can be improved from measurements of the reionization optical depth and 21 cm observations of the reionization epoch. Such observations may therefore constrain the physics in the universe after recombination. In Chapter 8, I describe work in progress that will help to tighten such constraints even further.

1.2.1.2 Generation of primordial magnetic fields

In this section, I summarize the main mechanisms that may provide primordial magnetic fields in the early universe, starting from mechanisms available in the pre-recombination epoch and going backward to the QCD phase transition, the electroweak phase transition and inflation.

As pointed out by Harrison [1970], the presence of primordial vorticity before the recombination of hydrogen can lead to the generation of magnetic fields in this epoch. This is because in this era, Thomson scattering is much more effective for electrons than for protons. The electrons are therefore still effectively coupled to the radiation field and behave as relativistic matter, while the ions are already non-relativistic. The angular velocities of electrons and ions could therefore scale differently with redshift and give rise to an electric current that generates a magnetic field. In principle, this could lead to comoving field strenghts of up to ~ 10 nG on scales of 1 Mpc. However, as pointed out by Rebhan [1992], the amount of primordial vorticity is constrained by the requirement that it does not produce too large anisotropies in the CMB. This yields the following upper limit to a present-day field with coherence length L generated by this mechanism:

$$B_0(L) < 6 \times 10^{-18} h_{70}^{-2} \left(\frac{L}{1 \text{ Mpc}}\right)^{-3} \text{ G}, \quad (1.2)$$

where h_{70} is the Hubble constant normalized to 70 km s^{-1} Mpc^{-1}. As shown by Berezhiani & Dolgov [2004], such primordial vorticity could be due to photon diffusion in the second order temperature fluctuations. However, in this case strong fields are only obtained on small scales.

1.2 Open questions

As mentioned in Chapter 1.1.1, the QCD phase transition is a first-order phase transition taking place at temperatures of $T_{QCD} \sim 150$ MeV by bubble nucleation. Initially, these bubbles grow as burning deflagration fronts and release heat in the form of supersonic shock fronts in an out-of-equilibrium process. When the shock fronts collide, they reheat the plasma up to T_{QCD} and stop further bubble growth. From this point, the transition evolves in thermal equilibrium during the so-called co-existence phase. The latent heat from these bubbles compensates for the cooling from the expansion and keeps the temperature at T_{QCD}, until expansion wins and the remaining quark-gluon plasma pockets are hadronized.

According to Quashnock et al. [1989], magnetogenesis during the QCD phase transition proceeds via the formation of an electric field behind the shock fronts of the expanding bubbles. This is a consequence of the baryon assymetry, which is typically assumed to be present at this evolutionary stage. Quashnock et al. [1989] then expect a typical field strength of $B_l = 5$ G on a scale of 100 cm. The magnetic field on larger scales can be estimated following an approach of Hogan [1983] by performing a proper volume average over a larger number of magnetic dipoles of size l randomly oriented in space. Such an average gives

$$B_L = B_l \left(\frac{l}{L}\right)^{3/2}. \tag{1.3}$$

On a comoving scale of $L \sim 1$ AU, one could therefore expect a field strength of 2×10^{-8} nG at present time.

Cheng & Olinto [1994] showed that stronger fields could be produced during the coexistence phase of the QCD phase transition. In this phase, a baryon excess builds up in front of the bubble wall. Simulations of Kurki-Suonio [1988] indicate that this effect might enhance the baryon density contrast by a few orders of magnitude. Even more relevant is that the thickness of this layer of $\sim 10^7$ fm is considerably larger than the microphysical QCD length scale of ~ 1 fm. This could give rise to a comoving field of 10^{-7} nG on a scale of 1 pc. Even stronger fields can be produced in the presence of hydrodynamic instabilities [Sigl et al., 1997]. Such instabilities can occur during phase transitions when the transport of latent heat is dominated by the fluid flow. While it is not fully clear if these instabilities can really develop, they appear plausible for typical parameters of the QCD phase transition. In this case, one could obtain comoving field strengths of ~ 1 nG on a scale of 100 kpc. Future constraints on primordial magnetic fields may therefore provide more insight in this early epoch of the universe.

As mentioned in Chapter 1.1.1, it is not fully clear whether the electroweak phase transition is a phase transition of first or second order. Here, I will mostly focus on the possibility of a first-order phase transition, as also required by the electroweak baryogenesis scenario [Riotto & Trodden, 1999]. For a second-order phase transition, one expects weaker magnetic fields which are unlikely to be astrophysically relevant [Grasso & Rubinstein, 2001]. Baym et al. [1996] suggest that strong magnetic fields can be generated in this epoch via a

1. INTRODUCTION

dynamo mechanism. For this scenario, seed fields are provided from random magnetic fluctuations which are always present on scales of the order of a thermal wavelength [Grasso & Rubinstein, 2001]. Once the universe cools below the critical temperature of $T_c \sim 100$ GeV, the Higgs field locally tunnels into a phase with broken symmetry. The tunneling gives rise to the formation of broken phase bubbles which expand and convert the false vacuum energy into kinetic energy. For a wide range of model parameters, the expansion gives rise to a supersonic shock wave ahead of the burning front and turbulence fully develops. On galactic scales of ~ 5 kpc, this gives rise to a field strength of $10^{-11} - 10^{-8}$ nG.

The numbers given above for the co-moving field strength assumed that the magnetic field evolves with $(1+z)^2$, such that the magnetic energy density scales as the energy density of radiation. This is however not necessarily the case. The evolution of magnetic fields generated at the electroweak or the QCD phase transition was calculated by Banerjee et al. [2004], finding that part of the magnetic energy may decay in a turbulent cascade, while in the presence of non-zero helicity, energy may also be transferred from smaller to larger scales. They find particularly promising results for magnetic fields from the QCD phase transition, expecting co-moving field strengths in the range of $10^{-2} - 2$ nG and typical coherence lengths of 10 kpc up to 1 Mpc. Such fields can be probed by the mechanisms discussed in the following chapters.

Finally, the inflationary epoch provides several important ingredients for the generation of primordial magnetic fields [Turner & Widrow, 1988]:

- It naturally produces effects on very large scales, even larger than the Hubble horizon.

- It provides the dynamical means to amplify waves on long wavelengths.

- The universe is not a good conductor during the inflationary epoch, thus the ratio of magnetic to radiation energy can increase.

- Classical fluctuations of massless fields with wavelengths smaller than the Hubble horizon can grow super-adiabatically, with their energy density decreasing only as $(1+z)^2$ rather than $(1+z)^4$.

There is however also one major obstacle for the generation of magnetic fields in this era. As shown by Parker [1968], for a conformally flat metric like the Robertson-Walker metric, the background gravitational field does not produce particles if the underlying theory is conformally invariant. As the classical theory of electrodynamics is conformally invariant, Turner & Widrow [1988] proposed three possibilities to overcome this obstacle. First, they considered to break conformal invariance explicitly by introducing a coupling of the form $RA_\mu A^\mu$ or $R_{\mu\nu}A^\mu A^\nu$, where R is the Ricci scalar, $R_{\mu\nu}$ is the Ricci tensor and A^μ the electromagnetic field. Such terms would give the photons an effective, time-dependent mass. They showed

1.2 Open questions

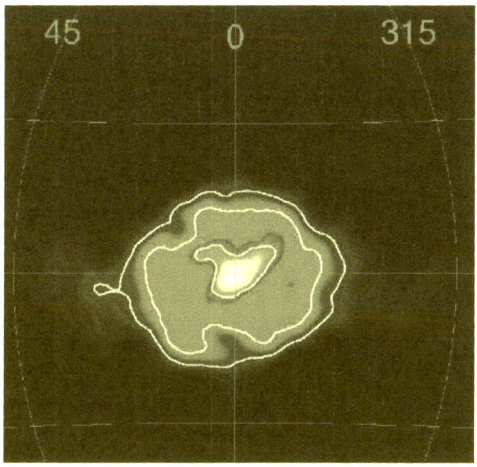

Figure 1.2: The observed 511 keV emission from the Galactic center: Evidence for dark matter annihilation? Credit: Ködlseder et al., A&A, 441, 513, 2003, reproduced with permission © ESO.

that such a mechanism could give rise to galactic magnetic fields without additional dynamo amplification.

Conformal invariance may also be broken by more complicated terms due to one-loop vacuum polarization effects in curved space-time. Unfortunately, Turner & Widrow [1988] showed that their contribution is far too small to be relevant. The third way they discussed to break conformal invariance involves a coupling of the photon to a charged field which is not conformally coupled, or the anomalous coupling to a pseudo-scalar. This possibility was considered in more detail by Ratra [1992] and Ratra & Peebles [1995], finding that the generated comoving field strength is highly model-dependent and ranges between 10^{-65} nG and 1 nG. More recent works still indicate enormous uncertainties regarding the magnetic field strength produced during inflation, including values with significant phenomenological consequences [Bamba et al., 2008; Campanelli et al., 2008; Campanelli, 2008].

1.2.2 Dark matter decay / annihilation scenarios

Particle physics models provide a number of different dark matter candidates. These include massive particles of ~ 100 GeV, like neutralinos or gravitinos that occur in supersymmetric

1. INTRODUCTION

extensions of the standard model. Very light axions of ~ 0.01 eV were suggested to solve the strong CP problem. Light dark matter particles in the MeV range have received particular attention due to the observation of 511 keV emission in the Galactic center. Kaluza-Klein particles are motivated in models with spatial extra-dimensions, and even more exotic candidates are considered in the literature. For the details of the different models, the interested reader is referred to recent reviews by Bertone et al. [2004]; Bergström [2000]; Ellis [2000].

It is interesting to note that many of these models predict some mechanism for annihilation or decay of dark matter particles, and weak interactions with baryons. Indeed, it is well-established that a thermally-averaged dark matter annihilation cross section of $\sim 3 \times 10^{-26}$ cm^3 s^{-1} naturally leads to the dark matter abundance observed in the universe. Conservative constraints from the Milky Way, Andromeda (M31), and the cosmic gamma-ray background yield upper limits on the annihilation cross section which are a few orders of magnitude larger, with the precise value depending on the dark matter mass [Mack et al., 2008]. The experimental upper limits for scattering cross sections between dark matter particles and baryons are 10^{-38} cm^2 for spin-dependent scattering [Desai et al., 2004; Angle et al., 2008] and 4×10^{-44} cm^2 for spin-independent scattering [Ahmed et al., 2008].

Observations of the Galactic center in different frequencies have provided further motivation for models that consider dark matter annihilation and/or decay. One finds an excess of GeV photons [de Boer et al., 2005], of microwave photons [Hooper et al., 2007], of positrons [Cirelli et al., 2008] and of MeV photons [Jean et al., 2006; Weidenspointner et al., 2006], which correlates with the Galactic bulge instead of the disk and cannot be attributed to single sources. It is therefore controversial whether it can be explained by conventional astrophysical sources alone or if dark matter annihilation models are required. It is not clear whether these different phenomena are related. Hence, their interpretation is still under discussion [de Boer, 2008]. As an example, the observed 511 keV emission from the Galactic center [Knödlseder et al., 2003] is shown in Fig. 1.2. Observations in the MeV energy range favor models based on light dark matter. These models suggest that light dark matter particles with masses of 1 – 100 MeV annihilate into electron-positron pairs [Boehm et al., 2004a]. In such a scenario, the dark matter mass needs to be larger than 511 keV in order to be able to produce electron-positron pairs through dark matter annihilation, however it should be smaller than 100 MeV, as otherwise pion final states that produce too many gamma rays would be possible [Boehm et al., 2004a]. Electromagnetic radiative corrections to the annihilation process require the emission of internal bremsstrahlung [Beacom et al., 2005]. It has also been proposed that such internal bremsstrahlung emission might explain the observed gamma-ray background in the 10 – 20 MeV range [Ahn et al., 2005; Ahn & Komatsu, 2005b] for dark matter masses of ~ 20 MeV, though these models appear less favorable in light of stronger upper limits on the dark matter particle mass [Beacom & Yüksel, 2006; Sizun et al., 2006]. Supernovae data require dark matter particle masses larger than 10 MeV, although this limit depends on assumptions made regarding the scattering cross section between dark matter particles and neutrinos [Fayet et al., 2006].

However, while light dark matter may explain the observed 511 keV emission, they have difficulties to explain the observations at GeV energies. For this purpose, a new model of so-called eXciting dark matter has been introduced by Finkbeiner & Weiner [2007]; Finkbeiner et al. [2008]. This model is based on massive dark matter candidates that can produce the required emission at high energies due to a Sommerfeld enhancement, and it assumes excited states for the dark matter particles, which can lead to emission of electron-positron pairs. At present, the FERMI satellite [1] searches for signals of dark matter annihilation both in the cosmic background and nearby galaxies, and may soon provide better constraints or even a detection of dark matter signals.

1.2.3 Implications for the first stars

Previous works often assumed that the initial conditions for the formation of the first stars are well-known, since the initial conditions for dark matter are well understood, the gas is not yet polluted with metals and no radiation fields are present apart from the cosmic microwave background. As a further simplification, the dynamical generation of magnetic fields during the collapse of the proto-cloud is mostly neglected. However, Tan & Blackman [2004] and Silk & Langer [2006] showed that an effective dynamo may be present in the protostellar disk, driving the magnetic field strength to equipartition with the thermal pressure. In the case where a magnetic field strength of 1 nG is already reached at densities of $\sim 10^3$ cm^{-3}, a protostellar jet could be launched which may blow away up to 10% of the accreting matter [Machida et al., 2006]. On the other hand, simulations of Price & Bate [2008] about present-day star formation in stellar clusters show that magnetic fields can indeed help to suppress fragmentation and help to form more massive stars. The effect of dynamically generated magnetic fields on the first stars is therefore not completely clear.

In the presence of primordial fields, the situation is however more complicated. Comoving field strengths of ~ 0.1 nG heat the IGM via ambipolar diffusion and decaying MHD turbulence, and therefore change the initial chemical conditions [Sethi et al., 2008]. The corresponding increase in the electron fraction enhances the formation of H_2 and HD, allowing the gas to cool almost down to the CMB floor, thus decreasing the thermal Jeans mass. On the other hand, the additional magnetic pressure can compensate for that and give rise to a very high magnetic Jeans mass. In the IGM, one finds

$$M_J^B \sim 10^{10} M_\odot \left(\frac{B_0}{3 \times 10^{-9} \text{ G}} \right)^3, \tag{1.4}$$

---

[1] http://www.nasa.gov/mission_pages/GLAST/main/index.html

1. INTRODUCTION

where B_0 is the comoving field strength. Primordial fields can thus prevent gas collapse in small minihalos, while in more massive halos, they may rather suppress fragmentation and support the formation of very massive stars. As the formation of stars in the presence of primordial magnetic fields is however not completely understood, we have explored the effect of different assumptions regarding the stellar population in Chapter 2. In collaboration with Daniele Galli, Simon Glover, Francesco Palla, Robi Banerjee, Raffaella Schneider and Ralf Klessen, I am currently investigating these effects in more detail and find that indeed primordial fields tend to stabilize a massive star formation mode.

The effect of dark matter annihilation on the first stars has been examined in more detail, but still with significant uncertainties remaining. As the first stars are expected to form on the dark matter cusps of the first minihalos, which are steepened further by adiabatic contraction during the formation of the stars, the annihilation of dark matter can be a strong energy source that may power the stars [Spolyar et al., 2008]. The additional energy input leads to an increase in the stellar radius, which may reach values up to ~ 1000 AU, thus reducing the surface temperature to ~ 3000 K because of the negative stellar heat capacity. This has been explored further by Iocco [2008] and Freese et al. [2008c], who considered the effect of scattering between baryons and dark matter particles, increasing the dark matter abundance in the star.

Iocco et al. [2008] considered dark star masses in the range $5 \leq M_* \leq 600\ M_\odot$ and calculated the evolution of the pre-main-sequence phase, finding that the dark star phase where the energy input from dark matter annihilation dominates may last for $10^2 - 10^4$ yr. Freese et al. [2008a] examined the formation process of the star in more detail, considering polytropic equilibria and additional mass accretion until the total Jeans mass of $\sim 800\ M_\odot$ is reached. They find that this process lasts for $\sim 5 \times 10^5$ yr. Iocco et al. [2008], Taoso et al. [2008] and Yoon et al. [2008] have calculated the stellar evolution for the case in which the dark matter density inside the star is enhanced by the capture of addition WIMPs via off-scattering from stellar baryons. Iocco et al. [2008] followed the stellar evolution until the end of He burning, Yoon et al. [2008] until the end of oxygen burning and Taoso et al. [2008] until the end of H burning. Yoon et al. [2008] also took the effects of rotation into account. The calculations found a potentially very long lifetime of dark stars and correspondingly a strong increase in the number of UV photons that may contribute to reionization.

1.3 Probing the early universe

As we have seen above, many open questions remain regarding the early universe and the epoch in which the first stars formed. The main goal of this thesis is to explore how the era between recombination and the end of reionization can be better understood. This concerns both the physics of the early universe, such as primordial magnetic fields and dark matter

1.3 Probing the early universe

annihilation / decay, the stellar population as well as the formation of supermassive black holes and the high metallicity in their host galaxies.

At present, we have only indirect information about the early universe, such as the observed reionization optical depth, the Gunn-Peterson troughs in quasars beyond $z = 6$ and the cosmic backgrounds, which can be used to constrain models of the early universe. However, we are currently approaching an era in which the early universe can be directly observed. This is possible for instance with 21 cm observations at high redshift with LOFAR [1], which are sensitive to the thermal gas properties on large scales. It may therefore observe the HII regions of the first quasars and the first galaxies, but also detected heating effects due to dark matter annihilation and decay. This possibility is discussed in more detail in Chapter 3.

The Planck satellite [2], which is planned to launch in April this year (2009), will measure the CMB with unprecedented accuracy, thus reducing the uncertainties regarding the cosmological parameters as well as the reionization optical depth. This may help to tighten the constraints presented in this work even further.

In about five years, the James Webb Space Telescope (JWST) will search for the first galaxies during the reionization epoch. At the same time, the Atacama Large Millimeter/submillimeter Array (ALMA) should be completed. As discussed in Chapter 7, the combination of these telescopes provides ideal conditions to explore the origin of the first quasars in the universe.

---

[1] http://www.lofar.org/
[2] http://www.rssd.esa.int/index.php?project=planck

1. INTRODUCTION

2

Reionization - A probe for the stellar population and the physics of the early universe[1]

In this chapter[1] , I discuss the possibility to constrain the stellar population and the physics of the early universe with the reionization optical depth as observed by WMAP [Schleicher et al., 2008b]. Particular emphasis is on primordial magnetic fields, dark matter annihilation and decay. Stellar populations consisting of very massive stars [Abel et al., 2002; Bromm & Larson, 2004], typical Pop. II stars [Scalo, 1998; Kroupa, 2002; Chabrier, 2003] and dark stars powered by the annihilation of dark matter [Spolyar et al., 2008; Iocco, 2008] are considered.

The following chapters then discuss further constraints on different scenarios in more detail. In particular, Chapter 3 examines the possibility to constrain or detect primordial magnetic fields with 21 cm observations, Chapter 5 considers further models for dark stars and the constraints from reionization and the cosmic backgrounds, and s-wave annihilation of light dark matter is ruled out in Chapter 6.

2.1 Introduction

Over the last decades, our understanding of cosmological reionization has improved considerably. Observations of high-redshift quasars clearly indicate that reionization must end

---

[1]Reprinted with permission from Schleicher, Banerjee & Klessen, PRD, 78, 083005, 2008. Copyright (2008) by the American Physical Society.

2. REIONIZATION - A PROBE FOR THE STELLAR POPULATION AND THE PHYSICS OF THE EARLY UNIVERSE[1]

around $z \sim 6$ [Becker et al., 2001], and the WMAP 5-year-measurement finds a reionization optical depth of $\tau = 0.087 \pm 0.017$ [Komatsu et al., 2008; Nolta et al., 2009], yielding clear evidence that reionization started significantly earlier and is thus a continuous process. Reionization through miniquasars would produce a soft-X-ray background that is significantly higher than the observed background, and can thus be ruled out [Dijkstra et al., 2004; Salvaterra et al., 2005]. Simulations of stellar reionization further show that the process is highly inhomogeneous and based on the growth and merging of ionized bubbles around the first stellar sources [Gnedin & Hui, 1998; Gnedin, 2000; Ciardi et al., 2003; Mellema et al., 2006]. Understanding the essential physical processes, it is possible to build semi-analytic models that describe stellar reionization, which can be tested over a wide parameter space [Shapiro & Giroux, 1987; Haiman & Loeb, 1997a; Barkana & Loeb, 2001; Loeb & Barkana, 2001; Choudhury & Ferrara, 2005; Schneider et al., 2006]. Additional constraints on the cosmic star formation rate are now available from gamma-ray burst observations [Yüksel et al., 2008]. The reionization framework thus provides an increasingly reliable test for the stellar population during reionization and can be used to constrain global physical conditions in the early universe.

Regarding the stellar population, Abel et al. [2002] and Bromm & Larson [2004] suggested that the first stars were top-heavy with a peak in the IMF at around 100 $M_\odot$. On the contrary, Clark et al. [2008] and Omukai et al. [2008] indicate that gas especially in more massive systems can fragment because of dips in the equation of state, which may lead to the formation of a stellar cluster. It was shown that cooling in previously ionized gas is enhanced and leads to typical stellar masses of $\sim 10\ M_\odot$ [Yoshida et al., 2007a,b]. It was further suggested that the presence of weak magnetic fields is sufficient to lead to a more present-day like mode of star formation, resulting in considerably lower masses [Silk & Langer, 2006]. Recently, a new phase of stellar evolution was suggested in which the stars would be powered by dark matter annihilation instead of nuclear fusion [Spolyar et al., 2008]. Various follow-up works have explored such a scenarios in more detail. Studies by Iocco [2008]; Freese et al. [2008b] explored the main-sequence and pre-main-sequence phase of dark stars, and other works calculated the effect of dark matter capture by off-scattering from baryons [Iocco et al., 2008; Yoon et al., 2008; Freese et al., 2008a,c; Taoso et al., 2008]. While many predictions are still model-dependent, it has often been suggested that such stars may have typical masses of 800 $M_\odot$, giving rise to a very bright main-sequence phase, or may have much longer lifetimes due to modifications in the stellar evolution. In the end, all these suggestions must face the constraint that the stellar population must be able to provide the correct reionization optical depth.

2.1 Introduction

Complications may arise through the presence of additional physics that are typically not considered in standard reionization calculations and simulations on the first stars. Such possibilities include the presence of primordial magnetic fields, dark matter decay and dark matter annihilation. Magnetic fields have been observed on all scales in the universe, and recently, it has been demonstrated that they were present already in high-redshift galaxies [Bernet et al., 2008]. They are found in the interstellar gas as well as in the intergalactic medium [Beck et al., 1996; Beck, 2001; Carilli et al., 2002], but their origin is still unclear. There is a viable possibility that these fields have a primordial origin [Grasso & Rubinstein, 2001; Widrow, 2002]. So far, the most stringent constraints on the strength of these putative fields come from the measurements of the cosmic microwave background radiation (CMBR) and from big-bang nucleosynthesis (BBN) calculations. A homogeneous magnetic field would produce temperature anisotropies in the CMBR [Zeldovich & Novikov, 1983] whose maximal amplitudes are limited by the COBE satellite measurements which in turn limit the field strength, as measured today, to $B_0 \lesssim 3.5 \times 10^{-9}$ G [Barrow et al., 1997]. The presence of any primordial field would also alter the CMBR power spectrum by changing the characteristic velocities. With the sensitivity of the PLANCK satellite [1] one should be able to detect fields with present day strength of $B_0 > 5 \times 10^{-8}$ G [Adams et al., 1996]. So far, measurements by the WMAP satellite [2] are compatible with the absence of primordial magnetic fields.

Strong magnetic fields in the early universe can also change the abundance of relic ^{4}He and other light elements during the big bang nucleosynthesis [Matese & O'Connell, 1969; Greenstein, 1969]. To comply with observational limits on light element abundances these primordial fields must not exceed 10^{12} G at the time when the universe was $T = 5 \times 10^9$ K which corresponds to a present day field $B_0 \lesssim 3 \times 10^{-7}$ G [Greenstein, 1969].

Effects of primordial magnetic fields have already been considered by Kim et al. [1996], finding that density perturbations can be enhanced by the Lorentz force from tangled magnetic fields. Both the evolution of perturbations in the presence of magnetic fields as well as their effect on the thermodynamics via ambipolar diffusion heating and decaying MHD turbulence was considered by Sethi & Subramanian [2005]. Recent calculations of Tashiro & Sugiyama [2006] show that the enhancement of structure due to magnetic fields is pronounced at about $5 \times 10^6 \, M_\odot$, but becomes less effective on larger mass scales and appears as a subdominant contribution on the scale of the magnetic Jeans mass. Consequences for 21 cm observations have been explored as well [Tashiro et al., 2006; Schleicher et al., 2009a]. In this work, we examine the consequences for reionization in more detail and calculate the backreaction on structure formation according to the work of [Gnedin, 2000]. The WMAP 5 year data [Nolta et al., 2009] have measured the Thomson scattering optical depth from reionization and allow to constrain different reionization scenarios in the early universe.

---

[1] http://sci.esa.int/science-e/www/area/index.cfm?fareaid=17
[2] http://map.gsfc.nasa.gov/

2. REIONIZATION - A PROBE FOR THE STELLAR POPULATION AND THE PHYSICS OF THE EARLY UNIVERSE[1]

The nature of dark matter is still unclear, and consequences of various particle physics like massive neutrinos or axion decay have been explored early [e. g. Doroshkevich et al., 1989; Berezhiani et al., 1990]. Observational progress allowed to refine these studies and to explore such scenarios in the framework of ΛCDM cosmology [Chen & Kamionkowski, 2004; Hansen & Haiman, 2004; Avelino & Barbosa, 2004; Kasuya & Kawasaki, 2004; Pierpaoli, 2004; Bean et al., 2003; Padmanabhan & Finkbeiner, 2005]. Recently, the possibility was discussed to detect such effects using future 21 cm telescopes [Furlanetto et al., 2006b], and it was shown that the fraction of the energy absorbed into the IGM can be calculated in detail for specific models of dark matter decay and annihilation [Ripamonti et al., 2007a]. Indeed, secondary ionization through the decay / annihilation products can provide a way to ionize the IGM which is independent of the stellar contribution, whereas the additional heat input increases the Jeans mass and delays the formation of the first structures and affects the chemical initial conditions [Ripamonti et al., 2007b].

In this work, we show how reionization constrains the properties of the stellar population as well as some additional heat sources like primordial magnetic fields, dark matter annihilation and decay. In Chapter 2.2, we present our model for stellar reionization, which considers the IGM as a two-phase-medium consisting of ionized bubbles and overall neutral gas. Based on the thermal evolution, we self-consistently determine the minimal mass scale of halos which can collapse. In Chapter 2.3, we explain our treatment of primordial magnetic fields and show how they modify the thermal evolution. The treatment of dark matter annihilation and decay, as well as some implications for dark stars, are discussed in Chapter 2.4. The optical depth for different models is given in Chapter 2.5. Further discussion and outlook is given in Chapter 2.6.

2.2 Reionization in the early universe

The thermal and ionization history of the IGM between recombination and the end of reionization is determined by a number of different processes. At redshifts $z > 300$, Compton scattering of CMB photons couples the gas temperature T to the CMB temperature T_{rad}. At lower redshifts, this coupling is less efficient and the gas decouples from the CMB due to adiabatic expansion. This standard scenario can be altered if additional energy injection mechanisms are present. Additional heat input can lead to an earlier redshift of decoupling, whereas an increase in the ionized fraction, for instance due to secondary ionization from the decay or annihilation products of dark matter, tends to make Compton cooling more efficient. Once structure formation sets in, the thermal and ionization history is further influenced from X-rays produced in star forming regions and UV photons that escape from the first galaxies. We modified the RECFAST code [Seager et al., 1999, 2000] and included all these feedback processes as described below, modeling the IGM as a two-phase medium of

2.2 Reionization in the early universe

ionized and partially-ionized gas. In this picture, the fully-ionized gas refers to the gas in the HII regions of the first luminous sources, while the partially ionized gas describes gas that is not yet affected by UV feedback.

2.2.1 The RECFAST code

In the absence of additional energy injection mechanisms, recombination in the early universe and freeze-out of electrons was calculated with unprecedent accuracy, following the detailed level populations of hundreds of energy levels for H, He and He$^+$ and self-consistently calculating the radiation field [Seager et al., 2000]. They developed the RECFAST code [1], a simplified version of the multi-level calculations, based on an effective three-level model for the hydrogen atom. RECFAST is capable of fully reproducing the results of the more detailed calculation [Seager et al., 1999]. Both the detailed calculation and the RECFAST code were recently updated and improved by Wong et al. [2008]. For the partially-ionized gas, we modify the equations describing the thermal and ionization history in the following way: The equation for the temperature evolution is given as

$$\frac{dT}{dz} = \frac{8\sigma_T a_R T_{\text{rad}}^4}{3H(z)(1+z)m_e c} \frac{x_e(T - T_{\text{rad}})}{1 + f_{\text{He}} + x_e} \\ + \frac{2T}{1+z} - \frac{2(L_{\text{heat}} - L_{\text{cool}})}{3nk_B H(z)(1+z)}, \quad (2.1)$$

where L_{heat} is the new heating term (see Chapter 2.3.1, 2.3.2 and 2.2.3), L_{cool} the new cooling term including Lyman α cooling, bremsstrahlung cooling and recombination cooling using cooling functions of Anninos et al. [1997], σ_T is the Thomson scattering cross section, a_R the Stefan-Boltzmann radiation constant, m_e the electron mass, c the speed of light, k_B Boltzmann's constant, n the total number density, $x_e = n_e/n_H$ the electron fraction per hydrogen atom, $H(z)$ is the Hubble factor and f_{He} is the number ratio of He and H nuclei, which can be obtained as $f_{\text{He}} = Y_p/4(1 - Y_p)$ from the mass fraction Y_p of He with respect to the total baryonic mass. The evolution of the ionized fraction of hydrogen, x_p, is given as

$$\frac{dx_p}{dz} = \frac{[C(z)x_e x_p n_H \alpha_H - \beta_H(1-x_p)e^{-h_p \nu_{H,2s}/kT}]}{H(z)(1+z)[1 + K_H(\Lambda_H + \beta_H)n_H(1-x_p)]} \quad (2.2) \\ \times \ [1 + K_H \Lambda_H n_H(1-x_p)] - \frac{k_{\text{ion}} n_H x_p}{H(z)(1+z)} - f_{\text{ion}}.$$

In this equation, n_H is the number density of hydrogen atoms and ions, h_p Planck's constant, k_{ion} is the collisional-ionization coefficient [Abel et al., 1997], f_{ion} describes ionization from

---
[1] http://www.astro.ubc.ca/people/scott/recfast.html

2. REIONIZATION - A PROBE FOR THE STELLAR POPULATION AND THE PHYSICS OF THE EARLY UNIVERSE[1]

X-rays (see Chapter 2.2.3), and the parametrized case B recombination coefficient for atomic hydrogen α_H is given by

$$\alpha_H = F \times 10^{-13} \frac{at^b}{1 + ct^d} \text{ cm}^3 \text{ s}^{-1} \tag{2.3}$$

with $a = 4.309$, $b = -0.6166$, $c = 0.6703$, $d = 0.5300$ and $t = T/10^4$ K, which is a fit given by Pequignot et al. [1991] to the coefficient of Hummer [1994]. This coefficient takes into account that direct recombination into the ground state does not lead to a net increase of neutral hydrogen atoms, since the photon emitted in the recombination process can ionize other hydrogen atoms in the neighbourhood. The fudge factor $F = 1.14$ serves to speed up recombination and is determined from comparison with the multilevel-code. We further introduce the clumping factor $C(z) \equiv \langle n_e^2 \rangle / \langle n_e \rangle^2$ to take into account the increase in the recombination rate in structures of increased density at low redshifts. We use the fit formula

$$C(z) = 27.466\exp(-0.114z + 0.001328z^2) \tag{2.4}$$

obtained from simulations of Mellema et al. [2006] at redshifts $z < 40$ and set $C(z) = 1$ at higher redshifts.

The photoionization coefficient β_H is calculated from detailed balance at high redshifts as described by Seager et al. [1999, 2000]. At lower redshifts, however, radiative recombination is no longer balanced by photoionization in the presence of additional energy injection mechanisms like ambipolar diffusion heating. Once the ionized fraction drops below 98%, we thus calculate the photoionization coefficient directly from the photoionization cross section given by Sasaki & Takahara [1993]. The frequency $\nu_{H,2p}$ corresponds to the Lyman-α transition from the $2p$ state to the $1s$ state of the hydrogen atom. The two-photon transition between the states $2s$ and $1s$ is close to Lyman-α. Consequently we use the same frequency for both processes. Finally, $\Lambda_H = 8.22458$ s^{-1} is the two-photon rate for the transition $2s$-$1s$ according to Goldman [1989] and $K_H \equiv \lambda_{H,2p}^3/[8\pi H(z)]$ the cosmological redshifting of Lyman α photons. The additional terms for Eq. (2.1) and (2.3) that are needed to describe the effects of dark matter annihilation and decay will be discussed in Chapter 2.4.

The fully-ionized component in the HII regions is described with the volume filling factor Q_{H^+} that denotes the volume fraction of ionized hydrogen bubbles. It is needed to compute the reionization optical depth and takes UV feedback into account. It evolves as

$$\frac{dQ_{H^+}}{dz} = \frac{Q_{H^+}C(z)n_{e,H^+}\alpha_A}{H(z)(1+z)} + \frac{dn_{ph}/dz}{n_H} \tag{2.5}$$

[Shapiro & Kang, 1987; Haiman & Loeb, 1997a; Barkana & Loeb, 2001; Loeb & Barkana, 2001; Choudhury & Ferrara, 2005; Schneider et al., 2006]. The UV photon production rate dn_{ph}/dz will be described in Chapter 2.2.3 in more detail, n_{e,H^+} denotes the mean electron

2.2 Reionization in the early universe

number density in fully-ionized regions, n_H the neutral hydrogen density in regions that are still unaffected from UV photons and $\alpha_A = 4.2 \times 10^{-13} [T_{max}/10^4 \text{ K}]^{-0.7}$ cm^3/s is the case A recombination coefficient [Osterbrock, 1989] which we have chosen here, as recombinations will preferably occur in dense regions where the recombination photons are unlikely to escape into the IGM [Miralda-Escudé, 2003]. It is evaluated at the temperature with $T_{max} = \max(10^4 \text{ K}, T)$ to account for the fact that the ionized regions should be heated at least to 10^4 K. If heating via ambipolar diffusion and decaying MHD turbulence increased the gas temperature above this threshold, it is evaluated at the gas temperature instead. To compare the models with the observational constraints from WMAP, we calculate the Thomson scattering optical depth of free electrons, given as

$$\tau_T = \frac{n_{H,0} c}{H_0} \int_0^{z_{re}} x_{eff} \sigma_T \frac{(1+z)^2}{\sqrt{\Omega_\Lambda + \Omega_m (1+z)^3}} dz, \qquad (2.6)$$

where $n_{H,0}$ is the comoving number density of ionized and neutral hydrogen, Ω_Λ and Ω_m the usual cosmological density parameters, H_0 the Hubble constant and z_{re} is the redshift where reionization starts, which we define as the point where the effective ionized fraction $x_{eff} = Q_{H^+} + (1 - Q_{H^+}) x_p$ becomes larger than 6.5%. This point has been chosen after a comparison of the actual TE cross calibration spectra with those from reionization models with fixed optical depth, but the results are not sensitive to this choice. In summary, we adopt the following picture:

- The partially-ionized gas unaffected by UV feedback is modeled with Eqs (2.1), (2.3) that describe the evolution of its temperature and ionization degree.

- The fully-ionized gas is described by the volume-filling factor Q_{H^+} calculated from Eq. (2.5). Its temperature is given as the maximum of 10^4 K and the temperature of the partially-ionized gas.

- Recombination both in the partially- and fully-ionized gas is enhanced by the clumping factor given in Eq. (2.4).

- For the partially-ionized component, recombination is described using the case B recombination coefficient, appropriate for low-density gas far from the ionizing sources.

- For the fully-ionized gas, we use the case A recombination coefficient, assuming that most recombinations take place in high-density regions near the ionizing sources.

The implementation of different feedback mechanisms is described below in more detail.

2. REIONIZATION - A PROBE FOR THE STELLAR POPULATION AND THE PHYSICS OF THE EARLY UNIVERSE[1]

2.2.2 The generalized filtering mass

The universe becomes reionized due to stellar feedback. We assume here that the star formation rate (SFR) is proportional to the change in the fraction of collapsed halos f_{coll}. As shown by Press & Schechter [1974], f_{coll} is given as

$$f_{coll} = \mathrm{erfc}\left[\frac{\delta_c(z)}{\sqrt{2}\sigma(m_{min})}\right], \quad (2.7)$$

where

$$\sigma^2(m) = \int_0^\infty \frac{dk}{2\pi^2} k^2 P_{lin}(k) \left[\frac{3j_1(kR)}{kR}\right]^2, \quad (2.8)$$

and where m_{min} is the minimum mass of haloes that are able to collapse at a given redshift. Here, $j_1(x) = (\sin x - x \cos x)/x^2$ is the spherical Bessel function, R is related to the halo mass M_h via $M_h = 4\pi\rho R^3/3$, ρ is the mean density, $\delta_c = 1.69/D(z)$ the linearized density threshold for collapse in the spherical top-hat model and $D(z)$ the linear growth factor. In the absence of magnetic field, m_{min} is determined by the filtering mass [Gnedin & Hui, 1998; Gnedin, 2000]. As discussed in the introduction, tangled magnetic fields can potentially create more small scale structure via the Lorentz force. The calculations of Tashiro & Sugiyama [2006] show that this effect is pronounced in the minihalo regime. While they adopted a constant minimal collapse mass of $10^6 \, h^{-1} \, M_\odot$ which is independent of the magnetic field, we use the framework of Gnedin [2000] to take into account the change in the mass scale of halos that can form stars. Indeed, we find that the mass scale is changed significantly (see Fig. 2.1(d)), such that the additional small-scale structure from tangled magnetic fields does not contribute to star formation. We introduce the magnetic Jeans mass, [Sethi & Subramanian, 2005]

$$M_J^B \sim 10^{10} M_\odot \left(\frac{B_0}{3 \times 10^{-9} \, \mathrm{G}}\right)^3, \quad (2.9)$$

and the thermal Jeans mass,

$$M_J = 2M_\odot \left(\frac{c_s}{0.2 \, \mathrm{km/s}}\right)^3 \left(\frac{n}{10^3 \, \mathrm{cm}^{-3}}\right)^{-1/2}. \quad (2.10)$$

The filtering mass in the presence of magnetic fields is then given as

$$M_{F,B}^{2/3} = \frac{3}{a} \int_0^a da' M_g^{2/3}(a') \left[1 - \left(\frac{a'}{a}\right)^{1/2}\right], \quad (2.11)$$

where $a = 1/(1 + z)$ is the scale factor and $M_g = \max(M_J, M_J^B)$. The minimum halo mass to consider is given as $m_{min} = \max(M_{F,B}, m_{cool})$, where m_{cool} denotes the minimal halo mass for which the baryons can efficiently cool after collapse. We adopt here the fiducial value $m_{cool} = 10^5 \, M_\odot$ [Greif et al., 2008]. Similar estimates have been given by Yoshida et al. [2003], Barkana & Loeb [2001] and Mackey et al. [2003]. For the subsequent analysis, we furthermore assume that only a certain fraction f_* of the collapsing halo mass turns into stars.

2.2.3 Stellar feedback

As X-ray photons have long mean free paths, they can play an important role in the ionization and heating of the gas. Assuming that the local correlation between the SFR and the X-ray luminosity (from 0.2 – 10 keV) holds up to a renormalization factor f_X [see Furlanetto et al., 2006a; Grimm et al., 2003; Ranalli et al., 2003; Gilfanov et al., 2004; Glover & Brand, 2003], the X-ray heating function L_X is given as

$$L_X = 3.4 \times 10^{40} f_X \left(\frac{\text{SFR}}{1 \, M_\odot \, \text{yr}^{-1}} \right) \text{erg s}^{-1}. \qquad (2.12)$$

X-ray emission has two major sources, inverse-Compton scattering of CMB photons with relativistic electrons accelerated in supernovae, and high-mass X-ray binaries. The former may play an increasingly important role at high redshifts, as the CMB photons are more energetic at high redshifts [Oh, 2001]. In the early universe the factor $f_X \sim 0.5$ if 10^{51} erg are released per 100 $M_\odot$, corresponding to an overall efficiency of 5% [Koyama et al., 1995]. The abundance of X-ray binaries depends on metallicity and the stellar initial mass function, and could be especially large if very massive Pop. III stars dominate [Glover & Brand, 2003]. In the model presented here, we adopt $f_X \sim 0.5$ as a generic value and assume that the uncertainty can be ascribed to the star formation efficiency f_*. In general, X-rays lose their energy through three channels. A fraction $f_{X,h}$ goes into heating, a fraction $f_{X,ion}$ into ionization, and a fraction $f_{X,coll}$ into excitation. These parameters should not be confused with f_{ion} introduced in Eq. (2.3) and f_{coll} introduced in Eq. (2.7). We calculate them using the fit formulae of Shull & van Steenberg [1985]. We can thus write

$$\frac{2L_X}{3k_B n_H} = 10^3 \, \text{K} \, f_X \left(\frac{f_*}{0.1} \frac{f_{X,h}}{0.2} \frac{df_{coll}/dz}{0.01} \frac{1+z}{10} \right) H(z). \qquad (2.13)$$

The contribution from the X-rays to the term f_{ion} in Eq. (2.3) is then given as

$$f_{ion} \sim \frac{f_{X,ion}}{f_{X,h}} \frac{L_X}{13.6 \, \text{eV} \, H(z)(1+z)}. \qquad (2.14)$$

UV photons from stellar sources are in general absorbed locally, only a fraction f_{esc} that manages to escape from the first galaxies can contribute to the ionization of the IGM. Again,

2. REIONIZATION - A PROBE FOR THE STELLAR POPULATION AND THE PHYSICS OF THE EARLY UNIVERSE[1]

the production of ionizing photons is associated with the star formation rate [Furlanetto et al., 2006a], yielding a contribution

$$\frac{dn_{ph}/dz}{n_H} \sim \xi \frac{df_{coll}}{dz}, \quad (2.15)$$

where

$$\xi = A_{He} f_* f_{esc} N_{ion}, \quad (2.16)$$

with $A_{He} = 4/(4-3Y_p) = 1.22$ and N_{ion} is the number of ionizing photons per stellar baryon.

2.2.4 Models for the stellar population

The nature of the first stellar sources is still under discussion, and even the question whether the primordial IMF is top-heavy or closer to the locally measured IMF is not solved. While for instance Abel et al. [2002] and Bromm & Larson [2004] found a top-heavy IMF using adaptive-mesh refinement (AMR) or smoothed-particle hydrodynamics (SPH) simulations, it was shown by Clark et al. [2008] that primordial and low-metallicity gas can fragment if the evolution of the gas is followed further after the formation of the first clump, due to a dip in the equation of state. Based on similar arguments, Omukai et al. [2008] argued that even a small metallicity fraction can lead to fragmentation in the first protogalaxies. It was further suggested that magnetic fields can have a crucial influence on the primordial IMF [Silk & Langer, 2006]. Given these uncertainties, we discuss different models for the stellar population. Dark stars have also been suggested as some of the first luminous sources. We will discuss this possibility in more detail in Chapter 2.4.3.

In models A and B, we assume that reionization is solely due to metal-free massive Pop III stars. This situation can only be examined from a theoretical point of view. Most investigations of UV feedback from high-mass zero-metallicity stars indicate that ionizing photons can easily escape the star-forming halo and drive large HII regions into the IGM. This suggests that the escape fraction in high-redshift galaxies could be very high, of order unity, being much higher than in the present-day universe [for instance Dove et al., 2000; Ciardi et al., 2002; Fujita et al., 2003]. Detailed numerical simulations of Whalen et al. [2004] show that indeed the shock bounding HII regions of massive Pop. III stars can easily photo-evaporate the minihalo and lead to an escape fraction of one. However, the situation is not fully clear. There are other studies that suggest that the escape fraction could remain small [Wood & Loeb, 2000], depending on the mass of the collapsing halo. This is why we adopt two different models. Model A is an extreme case, in which we assume that the escape fraction is 100% independent of the halo mass. The more realistic case is probably model B, where we distinguish whether the virial temperature corresponding to the generalized filtering mass is smaller or larger than 10^4 K. If it is smaller, we still assume that the halo is easily photoevaporated. If it is larger, we assume that most of the mass is kept within the

model	population	N_{ion}	f_{esc}
A	III	40,000	1
B	III	40,000	1 $(T_{vir} < 10^4 \text{ K})$
			0.1 $(T_{vir} \geq 10^4 \text{ K})$
C	III/II	10,000	1 $(T_{vir} < 10^4 \text{ K})$
			0.1 $(T_{vir} \geq 10^4 \text{ K})$
D	II	4,000	0.06

Table 2.1: Summary of adopted stellar models. The first column gives the model name. The second column indicates the stellar populations that contribute to reionization. The third column gives the number of ionizing photons per baryon used in Eq (2.16). The fourth column lists the adopted escape fractions. Model A is a highly extreme case in which we assume all ionizing photons come from massive Pop III stars and can escape the star-forming halo. In model B photons can escape efficiently only from halos less massive than a corresponding virial temperature of 10^4 K, while for higher-mass halos, this fraction is reduced to 10%. In model C we assume that Pop III as well as Pop II stars contribute to reionization, and we adopt the same escape probabilities as in B. Model D is another extreme case which assumes that all ionizing photons come from low-mass Pop II stars, with escape fraction 6%.

halo and adopt an escape fraction $f_{esc} = 0.1$. The virial mass corresponding to 10^4 K is given as $M_c = 5 \times 10^7 M_\odot \left(\frac{10}{1+z}\right)^{3/2}$ [Oh & Haiman, 2002; Greif et al., 2008]. For model A and B, we adopt a total number of ionizing photons per baryon of $N_{ion} = 40,000$, following Bromm et al. [2001].

The heavy elements produced by the very first stars will gradually enrich the IGM. It is very likely that some contribution to reionization comes from low-metallicity Pop II stars as well. As these stars are expected to have lower masses than Pop III stars, we introduce model C, which corresponds to a stellar population of intermediate mass. It has $N_{ion} = 10,000$ and an escape fraction according to model B, suggesting that such stars may photo-evaporate only low-mass halos.

Finally, we consider in model D the extreme case of very rapid chemical enrichment and assume cosmic reionization is entirely driven by low-mass Pop II stars. We assume a stellar mass distribution similar to the local IMF in the Galactic halo [Scalo, 1998; Kroupa, 2002; Chabrier, 2003] with a metallicity of 1/20 of the solar value. This corresponds to $N_{ion} \sim 4,000$. We furthermore adopt escape fractions that are typical for local star forming galaxies. Many upper limits and a few detections have been found observationally [Hurwitz et al., 1997; Heckman et al., 2001; Deharveng et al., 2001; Bland-Hawthorn & Maloney, 1999, 2001], suggesting $f_{esc} \sim 0.06$. A detection from Lyman-break galaxies at $z \sim 3$ implied a higher fraction of 10% [Steidel et al., 2001], while more recent observations place upper limits of 5 – 10% or claim detections at even lower levels [Giallongo et al., 2002;

2. REIONIZATION - A PROBE FOR THE STELLAR POPULATION AND THE PHYSICS OF THE EARLY UNIVERSE[1]

Malkan et al., 2003; Fernández-Soto et al., 2003; Inoue et al., 2005]. As the relevant rates scale only linear in $f_{\rm esc}$, it seems reasonable to adopt $f_{\rm esc} = 0.06$ for our Pop II model.

As it is unlikely that cosmic reionization occurs instantaneously with the onset of Pop III star formation nor that chemical enrichment is so rapid that all ionizing photons come from metal-enriched Pop II stars, we adopt the intermediate scenario C with a mixed population as our fiducial model. All global cosmological parameters are chosen according to the WMAP 5 year data [Komatsu et al., 2008].

2.3 The effect of magnetic fields on the IGM

As discussed by Sethi & Subramanian [2005]; Sethi et al. [2008], magnetic fields can significantly alter the thermal evolution of the IGM via ambipolar diffusion heating and decaying MHD turbulence, which can have a significant influence on the filtering mass scale. We explain our treatment of these heating terms in this section.

2.3.1 Ambipolar diffusion heating

The presence of magnetic fields introduces two different contributions to the heating rate, one coming from ambipolar diffusion and one resulting from the decay of MHD turbulence. In the first case, the contribution can be calculated as [Sethi & Subramanian, 2005; Cowling, 1956]:

$$L_{\rm ambi} = \frac{\rho_n}{16\pi^2 \gamma \rho_b^2 \rho_i} \left| \left(\nabla \times \vec{B} \right) \times \vec{B} \right|^2. \tag{2.17}$$

Here, ρ_n, ρ_i and ρ_b are the mass densities of neutral hydrogen, ionized hydrogen and all baryons. The ion-neutral coupling coefficient for primordial gas is given as [Draine, 1980; Shang et al., 2002]

$$\gamma = \frac{\frac{1}{2} n_H \langle \sigma v \rangle_{\rm H^+, H} + \frac{4}{5} n_{\rm He} \langle \sigma v \rangle_{\rm H^+, He}}{m_H \left[n_{\rm H} + 4 n_{\rm He} \right]}, \tag{2.18}$$

where $n_{\rm He}$ is the number density of He and m_H the mass of the hydrogen atom. Collisions with electrons are neglected here, as their contribution is suppressed by a factor m_e/m_H. We adopt the zero drift velocity momentum transfer coefficients of Pinto & Galli [2008] for collisions of H$^+$ with H and He, which is a good approximation in the absence of shocks. They are given by

$$\langle \sigma v \rangle_{\rm H^+, H} = 0.649 T^{0.375} \times 10^{-9} \text{ cm}^3 \text{ s}^{-1}, \tag{2.19}$$

$$\langle \sigma v \rangle_{\rm H^+, He} = (1.424 + 7.438 \times 10^{-6} T) \tag{2.20}$$

$$- 6.734 \times 10^{-9} T^2) \times 10^{-9} \text{ cm}^3 \text{ s}^{-1}.$$

As the power spectrum for the magnetic field is unknown and eq. (2.17) cannot be solved exactly, we adopt a simple and intuitive approach to estimate the integral for a given average magnetic field B with coherence length L. The heating rate can then be evaluated as

$$L_{\text{ambi}} \sim \frac{\rho_n}{16\pi^2 \gamma \rho_b^2 \rho_i} \frac{B^4}{L^2}. \quad (2.21)$$

The coherence length L is in principle a free parameter that depends on the generation mechanism of the magnetic field. It is constrained through the fact that tangled magnetic fields are strongly damped by radiative viscosity in the pre-recombination universe on scales smaller than the Alfvén damping scale k_{max}^{-1} given by [Jedamzik et al., 1998; Subramanian & Barrow, 1998; Seshadri & Subramanian, 2001]

$$\begin{aligned} k_{\text{max}} &\sim 234 \text{ Mpc}^{-1} \left(\frac{B_0}{10^{-9} \text{ G}}\right)^{-1} \left(\frac{\Omega_m}{0.3}\right)^{1/4} \\ &\times \left(\frac{\Omega_b h^2}{0.02}\right)^{1/2} \left(\frac{h}{0.7}\right)^{1/4}, \end{aligned} \quad (2.22)$$

where $B_0 = B/(1+z)^2$ denotes the comoving magnetic field. In fact, we expect fluctuations to be present on all scales. As the heating term goes as L^{-2}, we thus adopt the minimal value $L = k_{\text{max}}^{-1}/(1+z)$.

2.3.2 Decaying MHD turbulence

For decaying MHD turbulence, we adopt the prescription of Sethi & Subramanian [2005],

$$L_{\text{decay}} = \frac{B_0(t)^2}{8\pi} \frac{3\tilde{\alpha}}{2} \frac{[\ln(1 + t_d/t_i)]^{\tilde{\alpha}} H(t)}{[\ln(1 + t_d/t_i) + \ln(t/t_i)]^{\tilde{\alpha}+1}}, \quad (2.23)$$

where t is the cosmological time at redshift z, t_d is the dynamical timescale, t_i the time where decay starts, i. e. after the recombination epoch when velocity perturbations are no longer damped by the large radiative viscosity, z_i is the corresponding redshift. For a power spectrum of the magnetic field strength with power-law index α, the parameter $\tilde{\alpha}$ is given as $\tilde{\alpha} = 2(\alpha + 3)/(\alpha + 5)$ [Olesen, 1997; Shiromizu, 1998; Christensson et al., 2001; Banerjee & Jedamzik, 2003]. In the generic case, we expect the power spectrum of the magnetic field to have a maximum at the scale of the coherence length, and the heat input by MHD decay should be determined from the positive slope corresponding to larger scales [Müller

2. REIONIZATION - A PROBE FOR THE STELLAR POPULATION AND THE PHYSICS OF THE EARLY UNIVERSE[1]

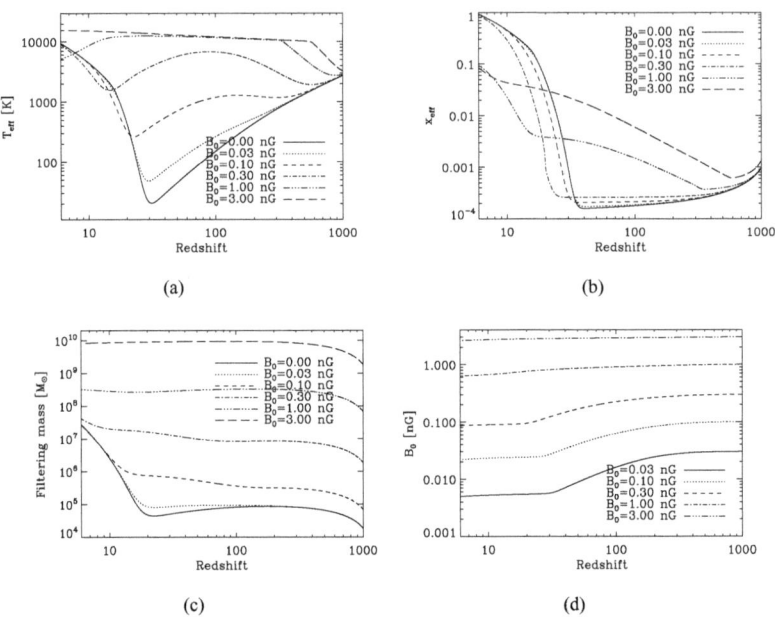

Figure 2.1: Evolution of different quantities as a function of redshift. a) The effective gas temperature. b) The effective ionized fraction. c) The filtering mass. d) The comoving magnetic field strength.

& Biskamp, 2000; Christensson et al., 2001; Banerjee & Jedamzik, 2003, 2004]. We thus adopt $\alpha = 3$ for the calculation. We estimate the dynamical timescale as $t_d = L/v_A$, where $v_A = B/\sqrt{4\pi\rho_b}$ is the Alvén velocity and ρ_b the baryon mass density. The evolution of the magnetic field as a function of redshift can be determined from the magnetic field energy $E_B = B^2/8\pi$, which evolves as [Sethi & Subramanian, 2005]

$$\frac{dE_B}{dt} = -4H(t)E_B - L_{\text{ambi}} - L_{\text{decay}}. \tag{2.24}$$

2.3.3 The evolution of the IGM

While the redshift of reionization of course depends on the model for the stellar population, the mechanism which delays reionization in the presence of magnetic fields is always the same. We illustrate this for our fiducal model C adopting a star formation efficiency f_* of 1%, but point out that the general discussion is valid also for the other stellar population models. We use the cosmological parameters of WMAP 5 [Komatsu et al., 2008]. Fig. 2.1(a) shows the thermal evolution of the IGM for different magnetic field strengths. In the absence of magnetic fields, the gas temperature follows the temperature of radiation until $z \sim 200$, where Compton scattering becomes inefficient. Afterwards, the gas cools adiabatically until it is reheated during reionization. In the presence of magnetic fields, gas and radiation decouple earlier due to the ambipolar diffusion heating and decaying MHD turbulence and stays at higher temperatures, which prevents collapse in low-mass halos and thus delays reionization. The additional heat increases the ionized fraction at early times (see Fig. 2.1(b)), while the redshift where the IGM becomes fully ionized is delayed, due to the increased generalized filtering mass (see Fig. 2.1(c)). Collisional ionization introduces a natural temperature plateau of the order ~ 10000 K, as any further temperature increase will have a backreaction on the ionized fraction of the gas, and thus make ambipolar diffusion less efficient.

2.4 Implications from the dark sector

A large variety of different particle physics models has been suggested to explain the dark matter content of the universe [Olive, 2008]. Many of these models predict some interactions in the dark sector and include some form of dark matter decay and annihilation. The consequences of such scenarios for the thermal evolution have been discussed in detail for instance by Ripamonti et al. [2007a,b]; Furlanetto et al. [2006b]. In this work, we discuss the implications of such scenarios on reionization. As we have seen for the case of primordial magnetic fields, the additional heat input increases the filtering mass and thus increases the minimal halo mass in which the first luminous objects can form, making stellar reionization less effective. On the other hand, once heating through dark matter decay or annihilation is effective, secondary ionization will likely be effective as well and thus increase the Thomson scattering optical depth. As a consequence of dark matter annihilation models, a new phase of stellar evolution was suggested by Spolyar et al. [2008]; Iocco [2008], in which stars are not powered by nuclear fusion, but by annihilating dark matter within them. These stars could reach masses of up to 10^3 $M_\odot$ [Freese et al., 2008a]. Once dark matter annihilation becomes ineffective and such stars enter the main sequence phase, they would thus be extremely bright sources of UV photons. The consequences of such a phase are discussed below. Further models taking into account a dark matter capturing phase have been suggested as well [Iocco et al., 2008; Yoon et al., 2008; Freese et al., 2008c,b]. As they are highly parameter-dependent, we study them in a separate work [Schleicher et al., 2008a].

2. REIONIZATION - A PROBE FOR THE STELLAR POPULATION AND THE PHYSICS OF THE EARLY UNIVERSE[1]

2.4.1 Dark Matter annihilation

The energy of the annihilation products can be deposited into heating, ionization and collisional excitation. The latter is quickly radiated away, but may contribute to the build-up of a Lyman α background that helps to couple the spin temperature of hydrogen to the gas temperature, which may be relevant for 21 cm observations [Furlanetto et al., 2006b]. For given models of dark matter, it is possible to work out the detailed absorbed fractions as a function of redshift [Ripamonti et al., 2007a]. In this work, however, we adopt the more generic approach of Furlanetto et al. [2006b]. For definiteness, they assume that the dark matter particles annihilate to high-energy photons, which allows to determine the energy fractions going into heat and ionization by the fitting formulae of Shull & van Steenberg [1985]. As they show, heating by dark matter annihilation is accompanied also by a significant increase in the electron fraction due to secondary ionization. For other models of dark matter annihilation, the results can be rescaled by appropriate absorption efficiencies. To Eq. (2.1) describing the temperature evolution, we thus add a term

$$\frac{\delta T}{\delta z} = -\frac{2}{3} \frac{\eta_2 m_p c^2}{\eta_1 k_B (1+z) H(z)} \xi_X \chi_h \qquad (2.25)$$

according to Furlanetto et al. [2006b], where $\eta_1 = 1 + f_{He} + x_i$, $\eta_2 = 1 + 4 f_{He}$, f_{He} is the helium fraction by number, χ_h is the fraction of energy going into heating [see Shull & van Steenberg, 1985] and ξ_X the effective baryon-normalized "lifetime", given as

$$\xi_X = \frac{\Omega_{DM} \rho_c^0}{m_{DM}} \langle \sigma v \rangle (1+z)^3 \left(\frac{\Omega_{DM}}{\Omega_b} \right). \qquad (2.26)$$

Here, Ω_{DM} and Ω_b denote the cosmological parameters for dark and baryonic matter, ρ_c^0 is the critical density at redshift zero, m_{DM} is the dark matter particle mass and $\langle \sigma v \rangle$ the velocity-averaged cross section. In the same way, we add a term to Eq. (2.3) that describes the evolution of the ionized fraction:

$$\frac{\delta x_i}{\delta z} = -\eta_2 \left(\frac{m_p c^2}{E_{ion}} \right) \xi_X \chi_i, \qquad (2.27)$$

where $E_{ion} = 13.6$ eV is the hydrogen ionization threshold and χ_i the fraction of the energy going into ionization, for which we use the fitting formulae of [Shull & van Steenberg, 1985]. As the decay rate scales with the dark matter density squared, it is most efficient at early times and may thus modify the recombination history, which allows to place upper limits on the dark matter annihilation. As shown by Zhang et al. [2006], the WMAP 1-year-data yield an

2.4 Implications from the dark sector

upper limit of

$$\langle\sigma v\rangle \leq 2.2 \times 10^{-29} \text{ cm}^3 \text{ s}^{-1} f_{abs}^{-1}\left(\frac{m_{DM}}{MeV}\right), \quad (2.28)$$

where f_{abs} corresponds to the energy fraction actually absorbed into the IGM and is of order 1 for particle masses in the MeV range [Ripamonti et al., 2007a].

2.4.2 Dark Matter decay

In a similar way, dark matter decay can alter the thermal and ionization history of the universe. The lifetimes of dark matter particles considered here will be considerably larger than the age of the universe to ensure that the abundance of dark matter does not change significantly. Still, the conversion of particle mass to thermal energy can have an important impact on the thermal evolution of the universe [Ripamonti et al., 2007a; Furlanetto et al., 2006b]. In this limit, the decay rate is constant over time and the effect at lower redshifts is more comprared to the case of dark matter annihilation. To calculate the heating and ionization rate, the same formalism can be employed as for dark matter annihilation, but we adopt the baryon-normalized decay rate

$$\xi_X = \frac{\Omega_{DM}}{\Omega_b t_X}, \quad (2.29)$$

where t_X is the lifetime of the dark matter particle. Based on the modified recombination and reionization histories in the presence of dark matter decay, it was shown that [Zhang et al., 2007]

$$\frac{f_\chi f_{abs}}{t_X} \leq 2.4 \times 10^{-25} \text{ s}^{-1}, \quad (2.30)$$

where f_χ is the fraction of the particle mass-energy released through decay and f_{abs} the fraction of the released energy deposited into the IGM. If data of large-scale surveys are included, the constraint can be slightly improved, yielding

$$\frac{f_\chi f_{abs}}{t_X} \leq 1.7 \times 10^{-25} \text{ s}^{-1}. \quad (2.31)$$

2.4.3 Dark stars

Dark stars have been suggested as a possible consequence of dark matter annihilation in the early universe [Spolyar et al., 2008; Iocco, 2008; Freese et al., 2008a]. If indeed the first stars form in the central peaks of the dark matter distribution in the early minihaloes, they might be powered by dark matter annihilation instead of nuclear fusion. Such stars could have up to 10^3 $M_\odot$ and surface temperatures in the range of 3000 – 10000 K, i. e.

2. REIONIZATION - A PROBE FOR THE STELLAR POPULATION AND THE PHYSICS OF THE EARLY UNIVERSE[1]

significantly colder than conventional Pop. III stars. However, this dark phase will gradually come to an end as the dark matter in the cusp annihilates, so the star may contract and finally reach a main sequence phase. A linear stability analysis and 1D simulations including hydrodynamics and radiation indicates the stability of such objects [Gamgami, 2007]. In this case, one can expect a similar number of photons per stellar baryon as for conventional Pop. III stars, perhaps even higher by a factor of 2 [Bromm et al., 2001]. As it is crucial for these objects to form on the peak of the dark matter density, we expect them to form only in low-mass halos with a virial temperature below 10^4 K, as the so-called atomic cooling halos are very turbulent [Greif et al., 2008] and are more likely to form a stellar cluster instead of a single massive star [Clark et al., 2008]. In addition, such massive halos are more likely to have accreted material from previous metal enrichment [Tornatore et al., 2007], which may cause fragmentation as well [Omukai et al., 2008]. In previously ionized regions, the thermodynamics of collapse are significantly altered and simulations generally find lower-mass stars [Yoshida et al., 2007a,b], which might be an important limitation for dark stars as well.

2.5 Constraining the parameter space with WMAP 5

In this section, we show the results for the different stellar populations and in the context of different additional physics.

2.5.1 Stellar reionization with and without primordial magnetic fields

We first concentrate on the effect of primordial magnetic fields and compare the results of different models for the stellar population. We run our reionization model for a range of different star formation efficiencies and primordial magnetic fields and obtain the corresponding optical depth. To constrain the parameter space, some assumptions need to be made. The first is obvious: The calculated optical depth must agree with the optical depth measured by WMAP 5 at least within 3σ. In addition, we require that the star formation efficiency may not be unreasonably high. For Pop. III stars with $\sim 100\ M_\odot$ forming in halos of $\sim 10^6\ M_\odot$ (dark matter), star formation efficiencies of the order 0.1% are expected. It seems thus reasonable to reject star formation efficiencies higher by one order of magnitude, i. e. of the order 1%. Further requirements are that reionization must be complete by redshift 6, and that there should be a stellar contribution to reionization, so that the universe becomes metal-enriched. Now we discuss the constraints resulting from these criteria.

2.5 Constraining the parameter space with WMAP 5

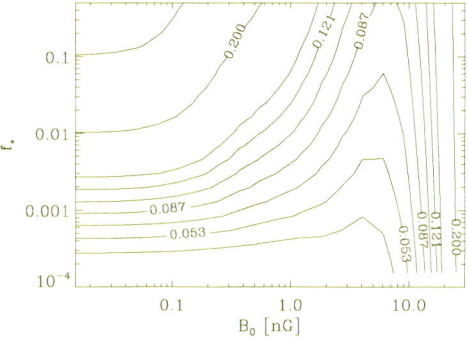

Figure 2.2: The reionization optical depth for Pop. III stars (model A) in the presence of primordial magnetic fields. The contour lines are equally spaced around $\tau_{re} = 0.87$ with $\Delta\tau = 0.017$, corresponding to the different σ-errors of the measurement. Magnetic fields higher than 20 nG are clearly excluded by the optical depth alone. The transition between stellar reionization and collisional ionization occurs at about 5 nG, providing a more stringent limit if metal enrichment is required.

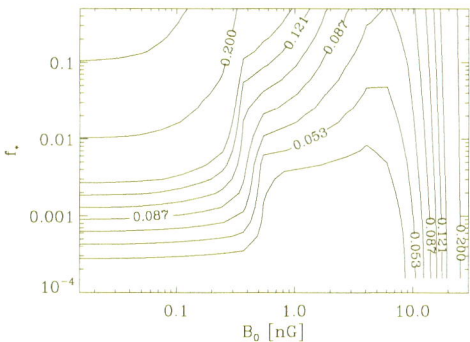

Figure 2.3: The reionization optical depth for Pop. III stars (model B) in the presence of primordial magnetic fields. Magnetic fields higher than 20 nG are clearly excluded by the optical depth alone. The transition between stellar reionization and collisional ionization occurs at about 5 nG. We can further exclude magnetic fields between 2 and 5 nG within 3-σ, as too high star formation efficiencies would be required to obtained the measured optical depth.

2. REIONIZATION - A PROBE FOR THE STELLAR POPULATION AND THE PHYSICS OF THE EARLY UNIVERSE[1]

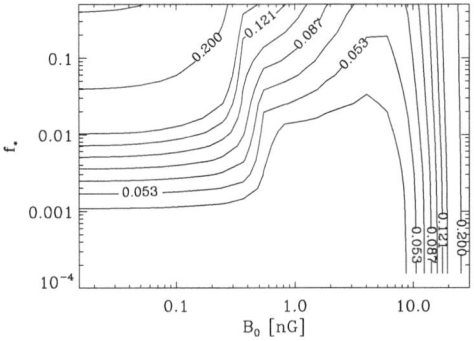

Figure 2.4: The reionization optical depth for a mixed population (model C) in the presence of primordial magnetic fields. Magnetic fields higher than 20 nG are clearly excluded by the optical depth alone. The transition between stellar reionization and collisional ionization occurs at about 5 nG. In addition, magnetic fields between 0.7 and 5 nG can be excluded, as they would require unreasonably high star formation efficiencies for the measured optical depth.

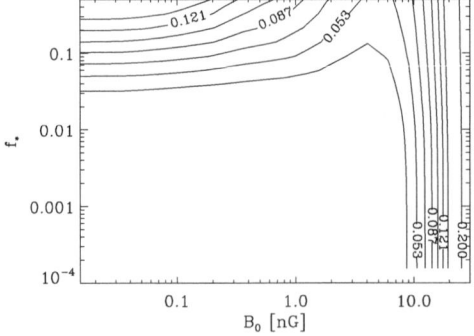

Figure 2.5: The reionization optical depth for Pop. II stars (model D) in the presence of primordial magnetic fields. For magnetic fields lower than 5 nG, stellar reionization dominates, but unreasonably high star formation efficiencies would be required to reconcile the measured optical depth. For stronger magnetic fields, collisional reionization dominates, which can be excluded if metal-enrichment is imposed. Magnetic fields larger than 20 nG can be excluded from the optical depth alone.

For all stellar models, we can exclude magnetic fields larger than 20 nG, based on the optical depth alone (see Fig. 2.2 to 2.5). a transition occurs at about 5 nG: For lower magnetic fields, reionization is still mostly due to stellar radiation, whereas for higher magnetic fields, the IGM is heated to such high temperatures that collisional ionization becomes very efficient and increases the optical depth. In this regime, structure formation is strongly suppressed, as the magnetic Jeans mass scales with B_0^3 and dominates over the thermal Jeans mass. Independently of the stellar model, such a regime is always found between 5 and 20 nG. Therefore, structure formation is considerably impeded such that metals do not form and the universe is not fully ionized by redshift 6. We can thus exclude magnetic field strength above 5 nG independent of the stellar model.

If we further require that the star formation efficiency must be lower than 1%, the results become model-dependent. In model A, which is an extreme case where very high escape fractions are assumed even for very massive halos, no further constraint on the magnetic field is possible on the 3σ level. Still, even in this extreme case, one can reject fields larger than 2.5 nG on the 1σ level. The other extreme case, model D with low escape fractions and a low number of ionizing photons per stellar baryon, can be rejected completely, as it always requires unphysically high star formation efficiencies.

For more realistic cases, essentially model B and C, we find that a critical magnetic field strength exists above which a very high star formation efficiency is needed to get into the 3σ interval around the measured optical depth. For model B, the critical value is 2 nG, for model C, it is at 0.7 nG. As it seems likely that a transition to less massive Pop. II stars occurs during reionization, model C might indeed be the most realistic case and provide an upper limit of 0.7 nG.

2.5.2 Dark reionization scenarios

For the case of dark matter annihilation and decay, we will show only the results for the Pop. III star model B. Regarding the competition between stellar reionization and secondary ionization from high-energetic annihilation / decay products, this corresponds to a rather conservative choice if one assumes that stellar reionization should do the main contribution. Such a stellar population could be interpreted either as conventional Pop. III stars or as so-called dark stars in a main sequence phase. We find that dark matter annihilation starts to become important for $\langle \sigma v \rangle / m_{DM} \sim 10^{-33}$ cm^3/s/eV and completely dominates the stellar contribution for higher values (see Fig. 2.6). Values higher than 3×10^{-33} cm^3/s/eV are inconsistent with the WMAP 5-year data. The results for dark matter decay are given in Fig.

2. REIONIZATION - A PROBE FOR THE STELLAR POPULATION AND THE PHYSICS OF THE EARLY UNIVERSE[1]

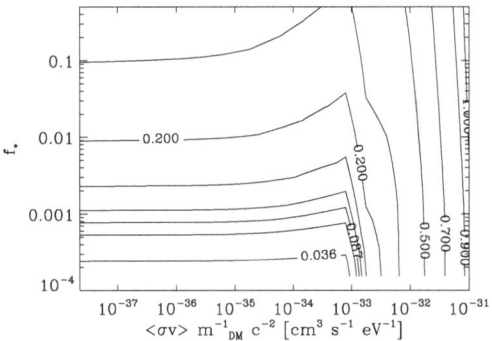

Figure 2.6: The reionization optical depth for Pop. III stars (model B) in the presence of dark matter annihilation. At $\langle\sigma v\rangle/m_{DM} \sim 10^{-33}$ cm^3/s/eV, secondary ionization of the annihilation products starts to dominate over stellar reionization. Values higher than 3×10^{-33} cm^3/s/eV are ruled out by the WMAP 5-year data.

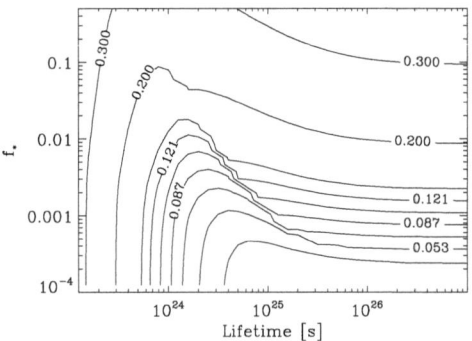

Figure 2.7: The reionization optical depth for Pop. III stars (model B) in the presence of dark matter decay. For lifetimes lower than 3×10^{24} s, secondary ionization of the annihilation products starts to dominate over stellar reionization. Lifetimes lower than 3×10^{23} s are ruled out by the WMAP 5-year data.

2.7. Decay starts to become important for lifetimes below 3×10^{24} s. However, lifetimes smaller than 3×10^{23} s are incompatible with WMAP 5. As pointed out recently, constraints on dark matter annihilation can be significantly improved taking into account the dark matter clumping factor $C(z) = \langle \rho_{DM}^2(z) \rangle / \langle \rho_{DM}(z) \rangle^2$ [Chuzhoy, 2008]. However, this quantity is highly uncertain and may vary by six orders of magnitude [Cumberbatch et al., 2008]. In the framework of light dark matter, we recently showed that is is constrained to be smaller than 10^5 at redshift zero [Schleicher et al., 2009b], yielding an enhancement of the annihilation rate by a factor of 10 or more at redshift 20. Other dark matter models even yield a clumping factor that drops down to 1 at these redshifts, such that our calculation altogether provides a conservative and firm upper limit [Cumberbatch et al., 2008].

From Fig. 2.6, we can further see that very massive dark stars in a main sequence phase, which should have higher star formation efficienes that are an order of magnitude higher than conventional Pop. III stars, are ruled out by WMAP 5, as the models require rather low star formation efficiencies to reproduce the Thomson scattering optical depth. Of course, this conclusion holds only if one assumes that reionization is completely due to dark stars, which seems rather unlikely (see discussion in Chapter 2.4.3), and we will discuss some alternatives in Chapter 2.6.

2.6 Conclusions and outlook

We have calculated the reionization optical depth for different stellar models in the presence of primordial magnetic fields as well as dark matter annihilation and decay. The results indicate which star formation efficiencies are required in the presence of some additional heating mechanism for a given stellar model. Considering different stellar models and primordial magnetic fields, we find the following results:

1. Independent of the model for the stellar population, we can securely exclude primordial magnetic fields larger than 5 nG.

2. For the most realistic case with a mixed stellar population (model C), we even find an upper limit of 0.7 nG, as higher magnetic fields would require star formation efficiencies larger than 1%, which is unrealistic. Similar results are found for model B, assuming reionization completely due to Pop. III stars.

3. Reionization only due to population II stars (model D) is ruled out completely.

For dark matter, we found the following results:

1. Dark matter annihilation provides noticeable contributions to the reionization optical depth only for thermally averaged mass-weighted cross sections $\langle \sigma v \rangle / m_{DM} \geq 10^{-33} \text{cm}^3/\text{s}/\text{eV}$.

2. REIONIZATION - A PROBE FOR THE STELLAR POPULATION AND THE PHYSICS OF THE EARLY UNIVERSE[1]

2. Parameters $\langle\sigma v\rangle/m_{DM} \geq 3 \times 10^{-33}$ cm^3/s/eV can be ruled out by 3σ on the basis of WMAP 5 year data.

3. Dark matter decay becomes important for the reionization optical depth for lifetimes below 3×10^{24} s.

4. Dark matter lifetimes below 3×10^{23} s are ruled out by 3σ.

These results are in agreement with conservative constraints obtained from the gamma-ray background [Mack et al., 2008]. We further showed that reionization can not be due to ~ 1000 $M_\odot$ dark stars alone, as the corresponding optical depth would be significantly too high. One might wonder whether heating from dark matter annihilation might help to significantly delay stellar reionization, in order to reconcile this model with observations. However, as can be seen in Fig. 2.6, this is more than compensated by the effects of secondary ionization, once that dark matter annihilation starts to have a significant influence on the IGM. In case collider experiments like the LHC [1] or other dark matter detection experiments [2] find evidence for a self-annihilating dark matter candidate, this can be seen as evidence for a rapid transition towards a different mode of star formation, or a problem in our understanding of dark stars. To reconcile dark stars with observations, the following scenarios seem feasible:

1. Dark stars with 1000 $M_\odot$ may be extremely rare objects, and their actual mass scale is closer to the mass scale for typical Pop. III stars. In such a case, reionization could not distinguish between dark stars and conventional Pop. III stars.

2. The transition to lower-mass stars occurs very rapidly, such that dark stars only contribute to the very early phase of reionization. Reasons for that might be chemical, radiative as well as mechanical feedback (see also the discussion in Chapter 2.4.3). In fact, even a double-reionization scenario might be conceivable, in which very massive dark stars ionize the universe at high redshifts. Due to chemical and radiative feedback, formation of such stars might be suppressed and the universe might become neutral again, until the formation of less massive stars becomes efficient enough to reionize the universe.

3. Dark stars do not reach a main-sequence phase, but are disrupted earlier by some nonlinear instability. Such an instability would be constrained by the fact that it neither appears in a linear stability analysis nore in 1D simulations. Violent explosions of dark stars might even be considered as a source for Gamma-Ray bursts.

---

[1] http://lhc.web.cern.ch/lhc/
[2] A list of dark matter detection experiments is given at http://cdms.physics.ucsb.edu/others/others.html.

2.6 Conclusions and outlook

While too definite conclusions on the existence of dark stars are not yet possible, it is at least indicated that the possibilities mentioned above should be explored in more detailed, and a better understanding of the properties of the dark stars. A better understanding of their evolution after the dark phase will certainly help to better understand their possible role during reionization.

The constraints derived here are independent of other works that essentially rely on the physics of recombination to derive upper limits on additional physics [e. g. Barrow et al., 1997; Zhang et al., 2006, 2007]. In particular, magnetic fields can evolve dynamically and their field strength may thus change between these epochs. It has recently been suggested that the Biermann Battery effect creates magnetic fields in the presence of an electron pressure gradient [Xu et al., 2008]. We thus need to probe magnetic fields at different epochs. As we will show in a separate work [Schleicher et al., 2009a], upcoming 21 cm measurements will allow to probe the thermal history before and during reionization in great detail, and may allow to detect primordial magnetic fields of the order of 0.1 nG.

Additional ways of probing the reionization history and the dark ages exist as well: Scattering of CMB-photons in fine-structure lines of heavy elements may lead to a frequency-dependent CMB power spectrum and may allow to measure metal-abundances as a function of redshift [Basu et al., 2004]. Before reionization, molecules may form in the IGM and introduce further frequency-dependent features in the CMB [Schleicher et al., 2008c]. Such features are likely enhanced in the presence of either primordial magnetic fields or dark matter decay / annihilation, as the increased electron fraction catalyses the formation of molecules, and the additional heat input leads to a departure of the level populations from the radiation temperature.

Further improvements are expected from the upcoming measurement of Planck, that will measure the reionization optical depth with unprecedent accuracy and thus allow to strengthen the constraints obtained here on the stellar populations and additional physics. In addition, a more accurate determination of the cosmological parameters will remove further uncertainties in the present models of reionization.

2. REIONIZATION - A PROBE FOR THE STELLAR POPULATION AND THE PHYSICS OF THE EARLY UNIVERSE[1]

3

Influence of primordial magnetic fields on 21 cm emission[1]

While the possibility to detect effects due to dark matter annihilation with 21 cm observations has been discussed in quite some detail in the literature [Furlanetto et al., 2006b; Chuzhoy, 2008; Cumberbatch et al., 2008], the signatures of primordial magnetic fields have been explored previously only for redshift $z > 30$ [Tashiro et al., 2006]. This is however difficult to observe. In particular, their model does not include the build-up of a Lyman α background by the first sources of light, which couples the excitation temperature of atomic hydrogen to the gas temperature and can lead to a net signal in 21 cm during the redshifts of reionization, where telescopes like LOFAR[1] will observe. In this chapter[2], I discuss why this may constrain primordial magnetic fields.

3.1 Introduction

Observations of the 21 cm fine structure line of atomic hydrogen have the potential to become an important means of studying the universe at early times, during and even before the epoch of reionization. This possibility was suggested originally by Purcell & Field [1956], and significant process in instrumentation and the development of radio telescopes has brought us close to the first observations from radio telescopes like LOFAR. While one of the main purposes is to increase our understanding of cosmological reionization [3], a number of further exciting applications have been suggested in the mean time. Loeb & Zaldarriaga [2004]

---

[1] http://www.lofar.org
[2] This chapter is based on an article by Schleicher, Banerjee & Klessen, ApJ, 692, 236, 2009, and reproduced with permission of the AAS.
[3] LOFAR Science Case: http://www.lofar.org/PDF/NL-CASE-1.0.pdf

3. INFLUENCE OF PRIMORDIAL MAGNETIC FIELDS ON 21 CM EMISSION[1]

demonstrated how 21 cm measurements can probe the thermal evolution of the IGM at a much earlier time, at redshifts of $z \sim 200$. Barkana & Loeb [2005a] suggested a method that separates physical and astrophysical effects and thus allows to probe the physics of the early universe. Furlanetto et al. [2006b] showed how the effect of dark matter annihilation and decay would be reflected in the 21 cm line, and effects of primordial magnetic fields have been considered by Tashiro et al. [2006].

Indeed, primordial magnetic fields can affect the early universe in various ways. The thermal evolution is significantly altered by ambipolar diffusion heating and decaying MHD turbulence [Sethi & Subramanian, 2005; Sethi et al., 2008; Schleicher et al., 2008b]. Kim et al. [1996] calculated the effect of the Lorentz force on structure formation and showed that additional power is present on small scales in the presence of primordial magnetic fields. It was thus suggested that reionization occurs earlier in the presence of primordial magnetic fields [Sethi & Subramanian, 2005; Tashiro & Sugiyama, 2006]. However, as pointed out by Gnedin & Hui [1998]; Gnedin [2000], the characteristic mass scale of star forming halos, the so-called filtering mass, increases signficantly when the temperature is increased. For comoving field strengths of ~ 1 nG, we found that the filtering mass scale is shifted to scales where the power spectrum is essentially independent of the magnetic field [Schleicher et al., 2008b]. Reionization is thus delayed in the presence of primordial magnetic fields. We further found upper limits of the order 1 nG, based on the Thomson scattering optical depth measured by WMAP 5 [Komatsu et al., 2008; Nolta et al., 2009] and the requirement that reionization ends at $z \sim 6$ [Becker et al., 2001].

The presence of primordial magnetic fields can have interesting implications on first star formation as well. In the absence of magnetic fields, Abel et al. [2002] and Bromm & Larson [2004] suggested that the first stars should be very massive, perhaps with ~ 100 solar masses. Clark et al. [2008] and Omukai et al. [2008] argued that in more massive and perhaps metal-enriched galaxies, fragmentation should be more effective and lead to the formation of rather low-mass stars, due to a stage of efficient cooling. Magnetic fields may change this picture and reduce the stellar mass by triggering jets and outflows [Silk & Langer, 2006]. However, simulations by [Machida et al., 2006] show that the change in mass is of the order 10%.

21 cm measurements can try to adress primordial magnetic fields in two ways: During the dark ages of the universe, at redshifts $z \sim 200$ well before the formation of the first stars, the spin temperature of hydrogen is coupled to the gas temperature via collisional de-excitation by hydrogen atoms [Allison & Dalgarno, 1969; Zygelman, 2005] and free electrons [Smith, 1966; Furlanetto & Furlanetto, 2007], constituting a probe at very early times. While collisional de-excitation becomes inefficient due to the expansion of the universe, the first stars

3.2 The evolution of the IGM

will build up a Lyman α background that will cause a deviation of the spin temperature from the radiation temperature by the Wouthuysen-Field effect [Wouthuysen, 1952; Field, 1958]. As primordial magnetic fields may shift the onset of reionization, the onset of this coupling constitutes an important probe on the presence of such fields. We adress these possibilities in the following way: In Chapter 3.2, we review our treatment of the IGM in the presence of primordial magnetic fields. The evolution of the 21 cm background and the role of Lyman α photons is discussed in Chapter 3.3. The evolution of linear perturbations in temperature and ionization is calculated in 3.4. The results for the power spectrum are given in Chapter 3.5, and the results are further discussed in Chapter 3.6.

3.2 The evolution of the IGM

As indicated in the introduction, primordial magnetic fields can have a strong impact on the evolution of the IGM before reionization. Once the first luminous objects form, their feedback must also be taken into account. In this section, we review the basic ingredients of our treatment of the IGM between recombination and reionization. We refer the interested reader to Schleicher et al. [2008b] for more details.

3.2.1 The RECFAST code

Our calculation is based on a modified version of the RECFAST code[1] [Seager et al., 1999, 2000] that calculates recombination and the freeze-out of electrons. The calculation of helium recombination was recently updated by Wong et al. [2008]. We have extended this code by including a model for reionization in the context of primordial magnetic fields Schleicher et al. [2008b]. The equation for the temperature evolution is given as

$$\begin{aligned} \frac{dT}{dz} &= \frac{8\sigma_T a_R T_r^4}{3H(z)(1+z)m_e c} \frac{x_e}{1+f_{\text{He}}+x_e}(T-T_r) \\ &+ \frac{2T}{1+z} - \frac{2(L_{\text{heat}} - L_{\text{cool}})}{3nk_B H(z)(1+z)}, \end{aligned} \quad (3.1)$$

where L_{heat} is the new heating term (see Chapter 3.2.2, 3.2.3 and 3.3.2), L_{cool} the cooling by Lyman α emission, bremsstrahlung and recombinations, σ_T is the Thomson scattering cross section, a_R the Stefan-Boltzmann radiation constant, m_e the electron mass, c the speed of light, k_B Boltzmann's constant, n the total number density, $x_e = n_e/n_H$ the electron fraction per hydrogen atom, $H(z)$ is the Hubble factor and f_{He} is the number ratio of He and H nuclei, which can be obtained as $f_{\text{He}} = Y_p/4(1-Y_p)$ from the mass fraction Y_p of He with respect to

---
[1] http://www.astro.ubc.ca/people/scott/recfast.html

3. INFLUENCE OF PRIMORDIAL MAGNETIC FIELDS ON 21 CM EMISSION[1]

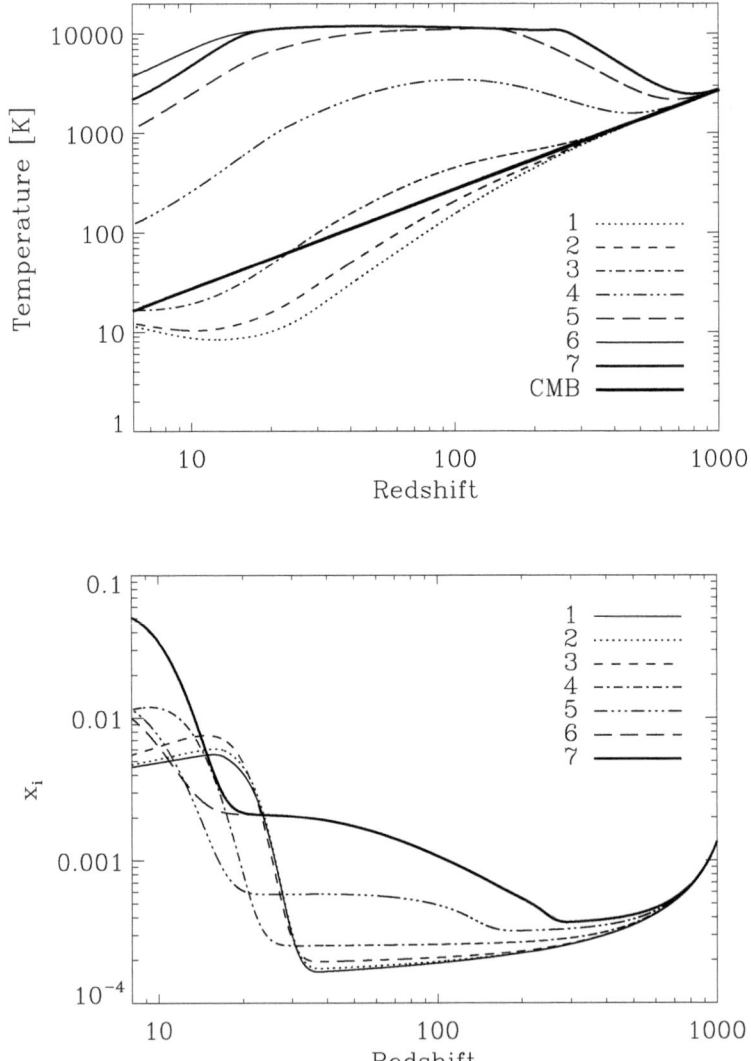

Figure 3.1: Evolution of gas temperature (upper panel) and ionized fraction (lower panel) in the medium that was not yet affected by UV feedback. The details of the models are given in Table 3.1.

3.2 The evolution of the IGM

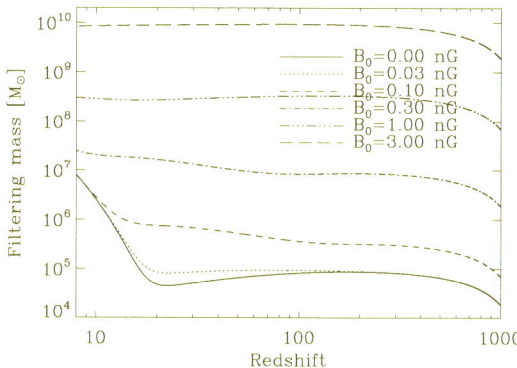

Figure 3.2: Evolution of the filtering mass scale for different comoving magnetic field strengths.

the total baryonic mass. Eq. (3.1) describes the change of temperature with redshift due to Compton scattering of CMB photons, expansion of the universe and additional heating and cooling terms that are described in the following subsections in more detail. The evolution of the ionized fraction of hydrogen, x_p, is given as

$$\begin{aligned}\frac{dx_p}{dz} &= \frac{[x_e x_p n_H \alpha_H - \beta_H(1-x_p)e^{-h_p \nu_{H,2s}/k_B T}]}{H(z)(1+z)[1+K_H(\Lambda_H+\beta_H)n_H(1-x_p)]} \\ &\times [1+K_H \Lambda_H n_H(1-x_p)] \\ &- \frac{k_{ion} n_H x_p}{H(z)(1+z)} - f_{ion},\end{aligned} \quad (3.2)$$

where the ionization fraction is determined by radiative recombination, photo-ionization of excited hydrogen atoms by CMB photons in an expanding universe and collisional ionization, as well as X-ray feedback. Here, n_H is the number density of hydrogen atoms and ions, h_p Planck's constant, f_{ion} describes ionization from X-rays and UV photons (see Chapter 3.2.3), and the parametrized case B recombination coefficient for atomic hydrogen α_H is given by

$$\alpha_H = F \times 10^{-13} \frac{at^b}{1+ct^d} \text{ cm}^3 \text{ s}^{-1} \quad (3.3)$$

with $a = 4.309$, $b = -0.6166$, $c = 0.6703$, $d = 0.5300$ and $t = T/10^4$ K [Pequignot et al., 1991; Hummer, 1994] . This coefficient takes into account that direct recombination into

3. INFLUENCE OF PRIMORDIAL MAGNETIC FIELDS ON 21 CM EMISSION[1]

the ground state does not lead to a net increase of neutral hydrogen atoms, since the photon emitted in the recombination process can ionize other hydrogen atoms in the neighbourhood.

The fudge factor $F = 1.14$ serves to speed up recombination and is determined from comparison with the multilevel-code. The photoionization coefficient β_H is calculated from detailed balance at high redshifts as described by Seager et al. [1999, 2000]. Once the ionized fraction drops below 98%, it is instead calculated from the photoionization cross section given by Sasaki & Takahara [1993], as detailed balance may considerably overestimate the photoionization coefficient in case of additional energy input at low redshift. The wavelength $\lambda_{H,2p}$ corresponds to the Lyman-α transition from the $2p$ state to the $1s$ state of the hydrogen atom. The frequency for the two-photon transition between the states $2s$ and $1s$ is close to Lyman-α and is thus approximated by $\nu_{H,2s} = c/\lambda_{H,2p}$ (i.e., the same averaged wavelength is used). Finally, $\Lambda_H = 8.22458$ s^{-1} is the two-photon rate for the transition $2s$-$1s$ according to Goldman [1989] and $K_H \equiv \lambda_{H,2p}^3/[8\pi H(z)]$ the cosmological redshifting of Lyman α photons. To take into account UV feedback, which can not be treated as a background but creates locally ionized bubbles, we further calculate the volume-filling factor Q_{HII} of ionized hydrogen according to the prescriptions of [Shapiro & Giroux, 1987; Haiman & Loeb, 1997a; Barkana & Loeb, 2001; Loeb & Barkana, 2001; Choudhury & Ferrara, 2005; Schneider et al., 2006]. We assume that the temperature in the ionized medium is the maximum between 10^4 K and the temperature in the overall neutral gas. The latter is only relevant if primordial magnetic fields have heated the medium to temperatures above 10^4 K before it was ionized.

3.2.2 Heating due to primordial magnetic fields

The presence of magnetic fields leads to two different contributions to the heating rate, one coming from ambipolar diffusion and one resulting from the decay of MHD turbulence. In the first case, the contribution can be calculated as [Sethi & Subramanian, 2005; Cowling, 1956; Schleicher et al., 2008b]:

$$L_{\text{ambi}} \sim \frac{\rho_n}{16\pi^2 \gamma \rho_b^2 \rho_i} \frac{B^4}{L^2}. \qquad (3.4)$$

Here, ρ_n, ρ_i and ρ_b are the mass densities of neutral hydrogen, ionized hydrogen and all baryons. The ion-neutral coupling coefficient is calculated using the updated zero drift velocity momentum transfer coefficients of Pinto & Galli [2008] for collisions of H$^+$ with H and He. The coherence length L is estimated as the inverse of the Alfvén damping wavelength k_{max}^{-1} given by [Jedamzik et al., 1998; Subramanian & Barrow, 1998; Seshadri & Subramanian, 2001] as

$$k_{\text{max}} \sim 234 \text{ Mpc}^{-1} \left(\frac{B_0}{10^{-9} \text{ G}}\right)^{-1} \left(\frac{\Omega_m}{0.3}\right)^{1/4}$$

$$\times \left(\frac{\Omega_b h^2}{0.02}\right)^{1/2} \left(\frac{h}{0.7}\right)^{1/4}, \qquad (3.5)$$

which is the scale on which fluctuations in the magnetic field are damped out during recombination. The comoving magnetic field is denoted as $B_0 = B/(1+z)^2$.

For decaying MHD turbulence, we adopt the prescription of Sethi & Subramanian [2005],

$$L_{\text{decay}} = \frac{B_0(t)^2}{8\pi} \frac{3\tilde{\alpha}}{2} \frac{[\ln(1+t_d/t_i)]^{\tilde{\alpha}} H(t)}{[\ln(1+t_d/t_i) + \ln(t/t_i)]^{\tilde{\alpha}+1}}, \qquad (3.6)$$

where t is the cosmological time at redshift z, t_d is the dynamical timescale, t_i the time where decay starts, i. e. after the recombination epoch when velocity perturbations are no longer damped by the large radiative viscosity, z_i is the corresponding redshift. For a power spectrum of the magnetic field with power-law index α, implying that the magnetic field scales with the wavevector k to the power $3 + \alpha$, the parameter $\tilde{\alpha}$ is given as $\tilde{\alpha} = 2(\alpha + 3)/(\alpha + 5)$ [Olesen, 1997; Shiromizu, 1998; Christensson et al., 2001; Banerjee & Jedamzik, 2003]. In the generic case, we expect the power spectrum of the magnetic field to have a maximum at the scale of the coherence length, and the heat input by MHD decay should be determined from the positive slope corresponding to larger scales [Müller & Biskamp, 2000; Christensson et al., 2001; Banerjee & Jedamzik, 2003; Banerjee et al., 2004]. We thus adopt $\alpha = 3$ for the calculation. We estimate the dynamical timescale as $t_d = L/v_A$, where $v_A = B/\sqrt{4\pi\rho_b}$ is the Alvén velocity and ρ_b the baryon mass density. The evolution of the magnetic field as a function of redshift can be determined from the magnetic field energy density $E_B = B^2/8\pi$, which evolves as [Sethi & Subramanian, 2005]

$$\frac{dE_B}{dt} = -4H(t)E_B - L_{\text{ambi}} - L_{\text{decay}}. \qquad (3.7)$$

For the models given in Table 3.1, the evolution of temperature and ionization in the overall neutral medium, i. e. the gas that was not yet affected by UV photons from reionization, is given in Fig. 3.1.

For comoving fields weaker than 0.02 nG, the gas temperature is closely coupled to the CMB via Compton-scattering at redshifts $z > 200$. At lower redshifts, this coupling becomes inefficient and the gas cools adiabatically during expansion, until heating due to X-ray feedback becomes important near redshift 20. The ionized fraction drops rapidly from fully ionized at $z \sim 1100$ to $\sim 2 \times 10^{-4}$, until it increases again at low redshift due to X-ray feedback. For larger magnetic fields, the additional heat input becomes significant and allows the gas to decouple earlier from the CMB. For comoving fields of order 0.5 nG or more, the gas reaches a temperature plateau near 10^4 K. At this temperature scale, collisional ionizations

3. INFLUENCE OF PRIMORDIAL MAGNETIC FIELDS ON 21 CM EMISSION[1]

become important and the ionized fraction in the gas increases, such that ambipolar diffusion heating becomes inefficient. The onset of X-ray feedback is delayed to the higher filtering mass in the presence of magnetic fields, which is discussed in the next section.

3.2.3 The filtering mass scale and stellar feedback

The universe becomes reionized due to stellar feedback. We assume here that the star formation rate (SFR) is proportional to the change in the collapsed fraction f_{coll}, i. e. the fraction of mass in halos more massive than m_{min}. This fraction is given by the formalism of Press & Schechter [1974]. In Schleicher et al. [2008b], we have introduced the generalized filtering mass $m_{F,B}$, given as

$$M_{F,B}^{2/3} = \frac{3}{a} \int_0^a da' M_g^{2/3}(a') \left[1 - \left(\frac{a'}{a}\right)^{1/2} \right], \tag{3.8}$$

where $a = 1/(1+z)$ is the scale factor and M_g is the maximum of the thermal Jeans mass M_J and the magnetic Jeans mass M_J^B, the mass scale below which magnetic pressure gradients can counteract gravitational colapse [Sethi & Subramanian, 2005; Subramanian & Barrow, 1998]. The concept of the filtering mass, i. e. the halo mass for which the baryonic and dark matter evolution can decouple, goes back to Gnedin & Hui [1998], see also Gnedin [2000]. To take into account the back reaction of the photo-heated gas on structure formation, the Jeans mass is calculated from the effective temperature $T_{eff} = Q_{HII} T_{max} + (1 - Q_{HII})T$, where $T_{max} = \max(10^4 \text{ K}, T)$ [Schneider et al., 2006]. The evolution of the filtering mass for different magnetic field strengths is given in Fig. 3.2.

When the comoving field is weaker than 0.03 nG, the filtering mass is of order 10^5 $M_\odot$ at high redshift and increases up to 10^7 $M_\odot$ during reionization, as UV and X-ray feedback heats the gas temperature and prevents the collapse of baryonic gas on *small scales*. Stronger magnetic fields provide additional heating via ambipolar diffusion and decaying MHD turbulence, such that the filtering mass is higher from the beginning. This implies that the first luminous objects form later, as the gas in halos below the filtering mass cannot collapse. This effect is even stronger when the comoving field is stronger than 0.3 nG, as the magnetic Jeans mass then dominates over the thermal Jeans mass and increases the filtering mass scale by further orders of magnitude. As discussed in Schleicher et al. [2008b], the feedback on structure formation via the filtering mass can delay reionization significantly, which in turn allows to calculate upper limits on the magnetic field strength due to the measured reionization optical depth.

Consequently, we define the lower limit to halo masses that can form stars as $m_{min} = \max(M_{F,B}, 10^5 \ M_\odot)$, where 10^5 $M_\odot$ is the minimal mass scale for which baryons can cool efficiently, as found in simulations of Greif et al. [2008]. We take into account X-ray feedback that can penetrate into the IGM due to its long mean-free path, as well as feedback from UV photons produced in the star forming regions. As we showed in Schleicher et al.

[2008b], Population II stars can not reionize the universe efficiently. It is thus reasonable to assume that the first luminous sources were indeed massive Pop. III stars. While this may no longer be true once chemical and radiative feedback becomes efficient and leads to a different mode of star formation, such a transition seems not important for the main purpose of this work, which is to demonstrate how primordial magnetic fields shift the onset of reionization and the epoch where a Lyman α background builds up initially. We thus adopt model B of Schleicher et al. [2008b], which uses an escape fraction of 100% if the virial temperature is below 10^4 K, and 10% for the atomic cooling halos with higher virial temperature. It further assumes that 4×10^4 UV photons are emitted per stellar baryon during the lifetime of the star. This choice is in particular motivated by simulations of Whalen et al. [2004], that show that minihalos are easily photo-evaporated by UV feedback from Pop. III stars, yielding escape fractions close to 100%. A new study by Wise & Abel [2008] shows that the escape fraction may be larger than 25% even for atomic cooling halos. However, their study assumes purely primordial gas, which is unlikely in such systems, and we expect that studies which take into account metal enrichment will find lower escape fractions, perhaps of the order 10% as suggested here.

3.3 The 21 cm background

The 21 cm brightness temperature depends on the hyperfinestructure level populations of neutral hydrogen, which is described by the spin temperature $T_{\rm spin}$. In this section, we review the physical processes that determine the spin temperature, discuss the build-up of a Lyman-α background that can couple the spin temperature to the gas temperature at low redshift, and discuss various sources of 21 cm brightness fluctuations.

3.3.1 The spin temperature

The observed 21 cm brightness temperature fluctuation can then be conveniently expressed as [Furlanetto et al., 2006a]

$$\begin{aligned} \delta T_b &= 27 x_{\rm H}(1+\delta) \left(\frac{\Omega_b h^2}{0.023}\right) \left(\frac{0.15}{\Omega_m h^2} \frac{1+z}{10}\right)^{1/2} \\ &\times \left(\frac{T_S - T_r}{T_S}\right) \left(\frac{H(z)/(1+z)}{dv_\parallel/dr_\parallel}\right) \text{ mK}, \end{aligned} \quad (3.9)$$

where δ is the fractional overdensity, $x_{\rm H} = 1 - x_{\rm eff}$ the effective neutral fraction, $x_{\rm eff} = Q_{\rm H^+} + (1 - Q_{\rm H^+})x_{\rm p}$ the effective ionized fraction, Ω_b and Ω_m the cosmological density parameters

3. INFLUENCE OF PRIMORDIAL MAGNETIC FIELDS ON 21 CM EMISSION[1]

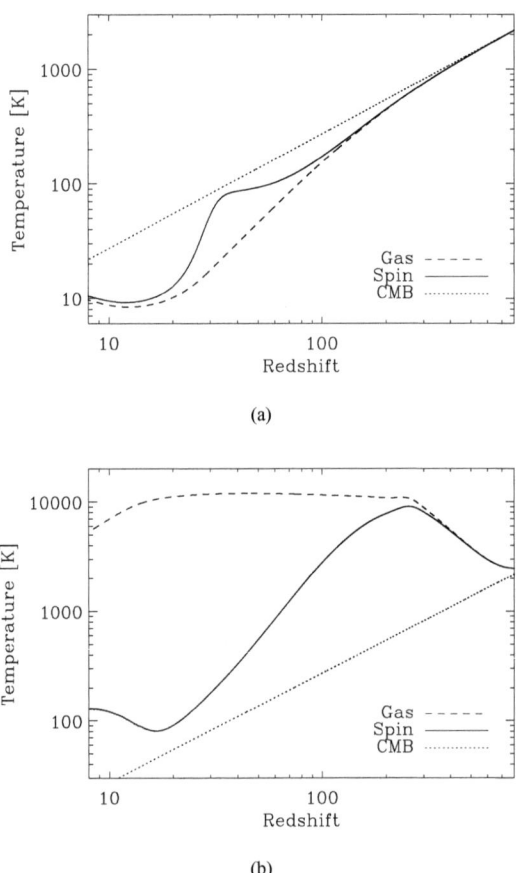

Figure 3.3: CMB, gas and spin temperature in the gas unaffected by reionization, for a star formation efficiency $f_* = 10^{-3}$, in the zero field case (upper panel) and for a field strength of 0.8 nG (lower panel). In the absence of magnetic fields, gas and CMB are strongly coupled due to efficient Compton scattering of CMB photons. At about redshift 200, the gas temperature decouples and evolves adiabatically, until it is reheated due to X-ray feedback. The spin temperature follows the gas temperature until redshift ~ 100. Collisional coupling then becomes less effective, but is replaced by the Wouthuysen-Field coupling at $z \sim 25$. In the presence of magnetic fields, additional heat goes into the gas, it thus decouples earlier and its temperature rises above the CMB. Wouthuysen-Field coupling becomes effective only at $z \sim 15$, as the high magnetic Jeans mass delays the formation of luminous sources significantly.

3.3 The 21 cm background

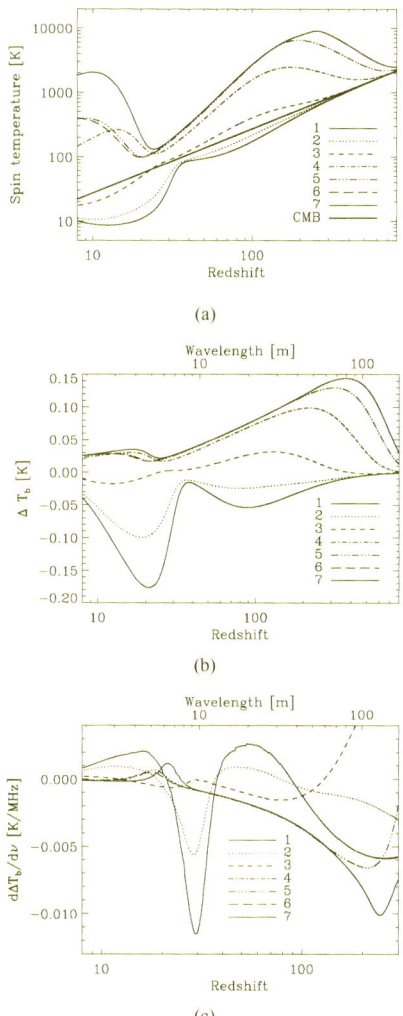

Figure 3.4: The mean evolution in the overall neutral gas, calculated for the models described in Table 3.1. Top: The spin temperature. Middle: The mean brightness temperature fluctuation. Bottom: The frequency gradient of the mean brightness temperature fluctuation.

3. INFLUENCE OF PRIMORDIAL MAGNETIC FIELDS ON 21 CM EMISSION[1]

for baryonic and total mass, and $dv_\parallel/dr_\parallel$ the gradient of the proper velocity along the line of sight, including the Hubble expansion. The spin temperature T_{spin} is given as.

$$T_{spin}^{-1} = \frac{T_r^{-1} + x_c T^{-1} + x_\alpha T_c^{-1}}{1 + x_c + x_\alpha}, \qquad (3.10)$$

where T_c is the colour temperature, $x_c = x_c^{HH} + x_c^{eH}$ the collisional coupling coefficient, given as the sum of the coupling coefficient with respect to hydrogen atoms, $x_c^{HH} = \kappa_{1-0}^H n_H T_*/A_{10}T_r$, and the coupling coefficient with respect to electrons, $x_c^{eH} = \kappa_{1-0}^e n_e T_*/A_{10}T_r$. Further, n_H denotes the number density of hydrogen atoms, $T_* = 0.068$ K the temperature corresponding to the 21 cm transition and $A_{10} = 2.85 \times 10^{-15}$ s^{-1} the Einstein coefficient for radiative de-excitation. For the collisional de-excitation rate through hydrogen, κ_{1-0}^H, we use the recent calculation of Zygelman [2005], and we use the new rate of Furlanetto & Furlanetto [2007] for the collisional de-excitation rate through electrons, κ_{1-0}^H. When both the spin and gas temperature are significantly larger than the spin exchange temperature $T_{se} = 0.4$ K, as it is usually the case, the colour temperature is given as

$$T_c = T\left(\frac{1 + T_{se}/T}{1 + T_{se}/T_{spin}}\right). \qquad (3.11)$$

Spin and colour temperature thus depend on each other and need to be solved for simultaneously. The Wouthuysen-Field coupling coefficient is given as

$$x_\alpha = S_\alpha \frac{J_\alpha}{J_c}, \qquad (3.12)$$

where $J_c = 1.165 \times 10^{-10}(1+z)/20$ cm^{-2} s^{-1} Hz^{-1} sr^{-1} and J_α is the proper Lyman α photon intensity, given as

$$J_\alpha(z) = \sum_{n=2}^{n_{max}} f_{rec}(n) \int_z^{z_{max}} dz' \frac{(1+z)^2}{4\pi} \frac{c}{H(z')} \epsilon(\nu'_n, z') \qquad (3.13)$$

[see also Barkana & Loeb, 2005a]. f_{rec} is the recycling probability of Lyman n photons [Pritchard & Furlanetto, 2006; Hirata, 2006], $H(z)$ the Hubble function at redshift z, c the speed of light, n_{max} the highest considered Lyman n resonance, z_{max} the highest redshift at which star formation might occur and $\epsilon(\nu, z)$ the comoving emissivity. We adopt $n_{max} = 30$ and $z_{max} = 40$. The suppression factor S_α can be calculated as [Chuzhoy & Shapiro, 2006; Furlanetto et al., 2006a]

$$S_\alpha = \exp\left(-0.803 T^{-2/3}\left(10^{-6}\tau_{GP}\right)^{1/3}\right) \qquad (3.14)$$

3.3 The 21 cm background

from the optical depth for Lyman α photons,

$$\tau_{\rm GP} \sim 3 \times 10^5 x_{HI} \left(\frac{1+z}{7}\right)^{3/2}. \tag{3.15}$$

Examples for the evolution of the gas, spin and CMB temperature in the zero-field case and for a magnetic field of 0.8 nG are given in Fig. 3.3.

3.3.2 The build-up of a Lyman-α background

In the dark ages, the universe is transparent to photons between the Lyman α line and the Lyman limit. Luminous sources can thus create a radiation background between these frequencies that is redshifted during the expansion of the universe. When a photon is shifted into a Lyman line, it is scattered and can couple the spin temperature to the gas temperature as described above. As shown by Hirata [2006] and Barkana & Loeb [2005a], a population of Pop. III stars can already produce a significant amount of Lyman α radiation around redshift 25. We assume that emission is proportional to the star formation rate, which is assumed to be proportional to the change in the collapsed fraction $f_{\rm coll}$. It thus depends on the evolution of the generalized filtering mass in the presence of magnetic fields. As discussed in Furlanetto et al. [2006a] and Barkana & Loeb [2005a], the comoving emissivity is given as

$$\epsilon(\nu, z) = f_* n_b \epsilon_b(\nu) \frac{df_{\rm coll}}{dt}, \tag{3.16}$$

where n_b is the comoving baryonic number density and $\epsilon_b(\nu)$ the number of photons produces in the frequency interval $\nu \pm d\nu/2$ per baryon incorporated in stars. The emissivity is approximated as a power law $\epsilon_b(\nu) \propto \nu^{\alpha_s - 1}$, and the parameters must be chosen according to the model for the stellar population. As discussed above, we assume that the first stars are massive Pop. III stars, and adopt the model of Barkana & Loeb [2005a] with $N_\alpha = 2670$ photons per stellar baryon between the Lyman α and the Lyman β line and a spectral index $\alpha_s = 1.29$

In addition, secondary effects from X-ray photons may lead to excitation of hydrogen atoms and the production of further Lyman α photons. We calculate the production of Lyman α photons from the X-ray background for consistency. The contribution from the X-rays is then given as [Chen & Miralda-Escudé, 2004; Chuzhoy et al., 2006; Furlanetto et al., 2006a]

$$\begin{aligned}\Delta x_\alpha^{X-\text{ray}} &= 0.05 S_\alpha f_X \left(\frac{f_{X,\text{coll}}}{1/3} \frac{f_*}{0.1} \frac{df_{\text{coll}}/dz}{0.01}\right) \\ &\times \left(\frac{1+z}{10}\right)^3. \end{aligned} \tag{3.17}$$

3. INFLUENCE OF PRIMORDIAL MAGNETIC FIELDS ON 21 CM EMISSION[1]

We adopt $f_X = 0.5$, and calculate $f_{X,\text{coll}}$ from the fit-formulae of Shull & van Steenberg [1985]. (Notation: Please recall that f_{coll} denotes the fraction of collapsed dark matter, while $f_{X,\text{coll}}$ denotes the fraction of absorbed X-ray energy which goes into X-rays. The parameter f_X provides an overall normalization for the correlation between the star formation rate and the X-ray luminosity [Koyama et al., 1995].) For the models described in Table 3.1, we have calculated the evolution of the spin temperature and the mean 21 cm brightness fluctuation. From an observational point of view, it might be more reasonable to focus on the frequency gradient of the mean 21 brightness [Schneider et al., 2008]. As the frequency dependence of the foreground is known, such an analysis may help to distinguish between the foreground and the actual signal. Results for the spin temperature, the mean 21 cm brightness fluctuation and its frequency gradient are shown in Fig. 3.4.

At redshifts above 200, the spin temperature is coupled closely to the gas temperature due to collisions with hydrogen atoms. At lower redshifts, the coupling becomes inefficient and the spin temperature evolves towards the CMB temperature, until structure formation sets in and the spin temperature is again coupled to the gas temperature via the Wouthuysen-Field effect. The point where this happens depends on the magnetic field strength, which influences the filtering mass and thus the mass scale of halos in which stars can form. Especially in the presence of strong magnetic fields, this delay in structure formation also makes the production of Lyman α photons less efficient and the departure from the CMB temperature is only weak.

This bevahiour is reflected in the evolution of the mean brightness temperature fluctuation. For weak magnetic fields, the gas temperature is colder than the CMB because of adiabatic expansion, so the 21 cm signal appears in absorption. At redshift 40 where the spin temperature approaches the CMB temperature, a maximum appears as absorption goes towards zero. Comoving fields of 0.05 nG add only little heat and essentially bring the gas and spin temperatures closer to the CMB, making the mean brightness temperature fluctuation smaller. For stronger fields, the 21 cm signal finally appears in emission. Due to very effective coupling and strong temperature differences at redshifts beyond 200, a pronounced peak appears there in emission. At redshifts between 20 and 10 which are more accessible to the next 21 cm telescopes, magnetic fields generally reduce the expected mean fluctuation, essentially due to the delay in the build-up of a Lyman α background.

The evolution of the frequency gradient in the mean brightness temperature fluctuation essentially shows strong minima and maxima where the mean brightness temperature fluctuation changes most significantly. In the case of weak fields, this peak is near redshift 30 when coupling via the Wouthuysen-Field effect becomes efficient. For stronger fields, this peak is shifted towards lower redshifts, thus providing a clear indication regarding the delay of reionization.

Scattering of Lyman α photons in the IGM does not only couple the spin temperature to the gas temperature, but it is also a potential source of heat. Its effect on the mean tem-

3.3 The 21 cm background

perature evolution has been studied by a number of authors [Madau et al., 1997; Chen & Miralda-Escudé, 2004; Chuzhoy & Shapiro, 2006; Ciardi et al., 2002]. While a detailed treatment requires to separately follow the mean temperature of hydrogen and deuterium, a good estimate can be obtained using the approach of Furlanetto et al. [2006a]; Furlanetto & Pritchard [2006], i. e.

$$\frac{2}{3}\frac{L_{\text{heat,Ly}\alpha}}{n_H k_B T H(z)} \sim \frac{0.8}{T^{4/3}} \frac{x_\alpha}{S_\alpha} \left(\frac{10}{1+z}\right). \tag{3.18}$$

For star formation efficiencies of the order 0.1%, as they are adopted throughout most of this work, Lyman α heating leads to a negligible temperature change of the order 1 K.

Model	B_0 [nG]	f_*
1	0	0.1%
2	0.02	0.1%
3	0.05	0.1%
4	0.2	0.1%
5	0.5	0.1%
6	0.8	0.1%
7	0.8	1%

Table 3.1: A list of models for different co-moving magnetic fields and star formation efficiencies, which are used in several figures for illustrational purposes. We give the comoving magnetic field B_0 and the star formation efficiency f_*. For illustration purposes, all models assume a population of massive Pop. III stars. The amount of Lyman α photons produced per stellar baryon would be larger by roughly a factor of 2 if we were to assume Pop. II stars. Assuming that the same amount of mass goes into stars, the coupling via the Wouthuysen-Field effect would start slightly earlier, but the delay due to magnetic fields is still more significant.

3.3.3 21 cm fluctuations

Fluctuations in the 21 cm brightness temperature can be caused by the relative density fluctuations δ, relative temperature fluctuations δ_T and relative fluctuations in the neutral fraction δ_H. Fluctuations in the Lyman α background can be an additional source of 21 cm fluctuations [Barkana & Loeb, 2005a], but a detailed treatment of these fluctuations is beyond the scope of this work. These fluctuations are only important in the early stage of the build-up of such a background. Once x_α is significantly larger than unity, the spin temperature is coupled closely to the gas temperature, and small fluctuations in the background will not affect the coupling. We adopt the treatment of Loeb & Zaldarriaga [2004] to calculate the fractional

3. INFLUENCE OF PRIMORDIAL MAGNETIC FIELDS ON 21 CM EMISSION[1]

21 cm brightness temperature perturbation, given as

$$\delta_{21}(\vec{k}) = (\beta + \mu^2)\delta + \beta_H \delta_H + \beta_T \delta_T, \quad (3.19)$$

where μ is the cosine of the angle between the wavevector $\vec{k}$ and the line of sight, with the expansion coefficients [Barkana & Loeb, 2005b; Furlanetto et al., 2006a]

$$\beta = 1 + \frac{x_c}{x_{\text{tot}}(1 + x_{\text{tot}})}, \quad (3.20)$$

$$\beta_H = 1 + \frac{x_c^{HH} - x_c^{eH}}{x_{\text{tot}}(1 + x_{\text{tot}})}, \quad (3.21)$$

$$\beta_T = \frac{T_r}{T - T_r} + \frac{x_c}{x_{\text{tot}}(1 + x_{\text{tot}})} \frac{d \ln x_c}{d \ln T}. \quad (3.22)$$

The coefficient x_{tot} is given as the sum of the collisional coupling coefficient x_c and the Wouthuysen-Field coupling coefficient x_α.

3.4 The evolution of linear perturbations in the dark ages

To determine the fluctuations in the 21 cm line, we need to calculate the evolution of linear perturbations during the dark ages. We introduce the relative perturbation of the magnetic field, δ_B, and the relative perturbation to the ionized fraction, δ_i, which is related to the relative perturbation in the neutral fraction, δ_H, by $\delta_i = -\delta_H(1 - x_i)/x_i$. In the general case, different density modes trigger independent temperature and ionization fluctuations. This introduces a non-trivial scale dependence [Barkana & Loeb, 2005c; Naoz & Barkana, 2005]. However, on the large scales which might ultimately be observeable, the growing density mode dominates [Bharadwaj & Ali, 2004]. The temperature and ionization fluctuations δ_T and δ_i are thus related to the density fluctuations δ by $\delta_T = g_T(z)\delta$ and $\delta_i = g_i(z)\delta$ with redshift-dependent coupling factors $g_i(z)$ and $g_T(z)$. The time evolution of g_T and g_i is well-known in the absence of magnetic fields [Naoz & Barkana, 2005; Bharadwaj & Ali, 2004] and has also been studied for the case of dark matter annihilation and decay [Furlanetto et al., 2006b]. We further introduce the relative fluctuation in the magnetic field, $\delta_B = g_B\delta$. To extend the previous analysis to the situation considered here, we must consider the effect of the new terms due to ambipolar diffusion heating and decaying MHD turbulence in Eq. (3.1). As it has a complicated dependence on different quantities, we will keep this discussion rather generic, so that it can also be applied to other situations as well.

Let us consider a source term of the form

$$\left(\frac{\delta T}{\delta z}\right)_{\text{source}} = f(\rho, T, x_i, B). \quad (3.23)$$

3.4 The evolution of linear perturbations in the dark ages

Here, $f(\rho, T, x_i, B)$ corresponds to the term describing ambipolar diffusion heating and decaying MHD turbulence in Eq. (3.1). When the quantities ρ, T, x_i and B are perturbed, it is straightforward to show that this introduces further terms for the evolution of the temperature perturbation, which are of the form

$$\begin{aligned}\frac{\delta\delta_T}{\delta z} &= T^{-1}(\delta\rho\frac{\partial f}{\partial \rho} + T\delta_T\frac{\partial f}{\partial T} + \delta_i x_i\frac{\partial f}{\partial x_i} \\ &+ \delta_B B\frac{\partial f}{\partial B} - f\delta_T).\end{aligned} \quad (3.24)$$

These new terms describe the change in the relative temperature fluctuation, depending on the additional source term $f(\rho, T, x_i, B)$ and its derivatives with respect to gas density, temperature, ionized fraction and magnetic field. The behaviour is dominated by the properties of the ambipolar diffusion heating term. The time evolution of the coupling factors g_T and g_i is thus given as

$$\begin{aligned}\frac{dg_T}{dz} &= \frac{g_T - 2/3}{1+z} + \frac{x_i}{\eta_1 t_\gamma}\frac{g_T T_r - g_i(T_r - T)}{T(1+z)H(z)} \\ &+ T^{-1}(\rho\frac{\partial f}{\partial \rho} + Tg_T\frac{\partial f}{\partial T} + g_i x_i\frac{\partial f}{\partial x_i} \\ &+ g_B B\frac{\partial f}{\partial B} - fg_T),\end{aligned} \quad (3.25)$$

$$\frac{dg_i}{dz} = \frac{g_i}{1+z} + \frac{\alpha_H x_i n_H(1 + g_i + \alpha'_H g_T)}{(1+z)H(z)}. \quad (3.26)$$

The first term in Eq. (3.25) describes the evolution towards an adiabatic state in the absence of heating or cooling mechanisms. The second term describes the interaction with the CMB via Compton scattering, which in general drives the gas towards an isothermal state, as the heat input per baryon is constant. The further terms are those derived in Eq. (3.24). In Eq. (3.26), the first term holds the relative fluctuation δ_i constant in the absence of recombinations, while the second term describes the effect of hydrogen recombination, which tends to drive the corresponding coupling factor g_i towards -1. Fluctuations in density and ionization are then anticorrelated.

This system must be closed with an additional assumption for the fluctuation in the magnetic field. The dominant mechanism for energy loss is ambipolar diffusion, which scales with ρ_b^{-2}. It seems thus unlikely that the magnetic field will be dissipated in regions of enhanced density. We rather assume that the magnetic flux is approximately conserved. In the case of dynamically weak field strength, the corresponding surface scales as $\rho^{-2/3}$, so the magnetic field scales as $\rho^{2/3}$, thus $g_B \sim 2/3$. Ambipolar diffusion, the dominant new heating term, scales as ρ^{-2}. To relate the change in energy to the change in temperature, it is further

3. INFLUENCE OF PRIMORDIAL MAGNETIC FIELDS ON 21 CM EMISSION[1]

multiplied by $3/(2nk_B)$, such that in this case $f \propto \rho^{-3}$. This behavior is not fully compensated by the other terms, thus heating is less effective in overdense regions, and g_T tends to drop below the adiabatic value of $2/3$. For the models given in Table 3.1, the evolution of g_T and g_i is given in Fig. 3.5.

In the zero or weak magnetic field case, the coupling factor g_T evolves from zero (efficient coupling to the CMB) towards $2/3$ (adiabatic), until X-ray feedback sets in. For comoving fields larger than 0.05 nG, there is an early phase around $z \sim 500$ where it evolves towards adiabatic behaviour, as the gas recombines more efficiently in these early times and higher densities, where the factor g_i is still of order -1. Thus, ambipolar diffusion in density enhancements is more efficient and effectively heats the gas at these redshifts. Below redshift 500, the relative fluctuation in ionization evolves towards zero, as the gas becomes thinner and hotter. That leads to a phase in which the relative temperature fluctuations evolves towards an isothermal state ($g_T = 0$). At low redshifts, X-ray feedback leads to a decrease of the coupling factor g_T. Its behaviour is no longer adiabatic because of this additional heat input. The increase in the mean ionized fraction due to secondary ionizations leads to a more negative g_i.

3.5 The 21 cm power spectrum

We can now calculate the 21 cm power spectrum in the presence of primordial magnetic fields from Eq. (3.19), leading to

$$P_{21}(k,\mu) = \delta T_b^2 (\beta' + \mu^2)^2 P_{\delta\delta}(k), \qquad (3.27)$$

where $P_{\delta\delta}(k)$ is the baryonic power spectrum and

$$\beta' = \beta + \beta_T g_T - \beta_H x_i g_i / (1 - x_i). \qquad (3.28)$$

For simplicity, we average over the μ dependence and define $P_{\delta\delta} = P_0/(1+z)^2$, such that

$$P_{21}(k) = \delta T_b^2 \left(\beta'^2 + \frac{2}{3}\beta' + \frac{1}{5} \right) \frac{P_0(k)}{(1+z)^2}. \qquad (3.29)$$

On the large scales considered here, the power spectrum of the 21 cm line is proportional to the baryonic power spectrum $P_0(k)$ for a given redshift, but the proportionality constant will evolve with redshift and is independent of the wavenumber in the linear approximation. The ratio P_{21}/P_0 is shown in Fig. 3.5. It essentially reflects the behaviour that was found earlier for the mean brightness temperature fluctuation. At high redshift, the amplitude of the 21 cm power spectrum is larger in the presence of additional heat from ambipolar diffusion and

3.5 The 21 cm power spectrum

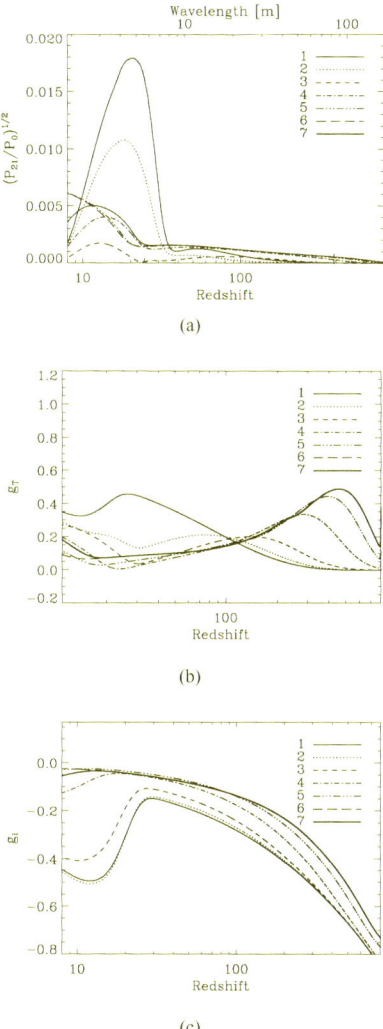

Figure 3.5: The evolution of the 21 cm large-scale power spectrum and the growth of large-scale fluctuations in temperature and ionization for the models given in Table 3.1. Top: The evolution of the 21 cm power spectrum, normalized with respect to the baryonic power spectrum at redshift 0. Middle: The evolution of the ratio $g_T = \delta_T/\delta$. Bottom: The evolution of the ratio $g_i = \delta_{x_i}/\delta$.

3. INFLUENCE OF PRIMORDIAL MAGNETIC FIELDS ON 21 CM EMISSION[1]

decaying MHD turbulence. At low redshifts, it is smaller in the presence of magnetic fields, as the high filtering mass delays the build-up of a Lyman α background and leads only to a weak coupling of the spin temperature to the gas temperature. In addition, the point where this coupling starts is shifted towards lower redshifts. This provides two signatures that can provide strong indications for the presence of primordial magnetic fields.

At high redshifts $z > 50$, the additional heat increases the difference between gas and radiation temperature, and the gas decouples earlier from radiation. All that tends to increase the 21 cm signal at high redshift. For redshifts $z < 40$, the situation is more complicated. The altered evolution of the fluctuations and the delay in the formation of luminous objects decrease the amplitude of the power spectrum in total, but also the onset of efficient coupling via the Wouthuysen-Field effect can be shifted significantly.

In this context, one might wonder whether 21 cm observations can actually probe additional small-scale power, which is predicted in the context of primordial magnetic fields [Kim et al., 1996; Sethi & Subramanian, 2005; Tashiro & Sugiyama, 2006]. However, as discussed for instance by Furlanetto et al. [2006a], the first 21 cm telescopes will focus on rather larger scales of the order 1 Mpc. The contribution to the power spectrum from primordial magnetic fields depends on the assumed power spectrum for the magnetic field. Regarding the formation of additional structures, the most significant case is a single-scale power spectrum. In this case, the contribution to the matter power spectrum can be evaluated analytically, yielding [Kim et al., 1996; Sethi & Subramanian, 2005]

$$P(k) = \frac{B_0^4 k^3 H_0^{-4}}{(8\pi)^3 \Omega_m^4 \rho_c^2 k_{max}^2} \tag{3.30}$$

for $k \leq 2k_{max}$, where k_{max} is given in Eq. 3.5. For interesting field strengths of the order of 1 nG, we thus expect modifications on scales of the order 1 kpc. For weaker fields, k_{max}^{-1} is shifted to larger scales, but the power spectrum decreases with B_0^4. Detecting such changes in the matter power spectrum is thus certainly challenging. However, the calculation of the filtering mass indicates that the baryonic gas will not collapse in minihalos of small masses when primordial fields are present. Star formation and HII regions from UV feedback are thus limited to more massive systems. This might be an additional way to distinguish the case with and without magnetic fields.

3.6 Discussion and outlook

In the previous sections, we have presented a semi-analytic model describing the post-reionization universe and reionization in the context of primordial magnetic fields, and calculated the consequences for the mean 21 cm brightness fluctuation and the large-scale power

3.6 Discussion and outlook

spectrum. formation of We identify two regimes in which primordial magnetic fields can influence effects measured with 21 cm telescopes. At low redshifts, primordial magnetic fields tend to delay reionization and the build-up of a Lyman α background, thus shifting the point where the signal is at its maximum, and changing the amplitude of the 21 cm power spectrum. The first 21 cm telescopes like LOFAR and others will focus mainly on the redshift of reionization and can thus probe the epoch when a significant Lyman α background builds up. As our understanding of the first stars increases due to advances in theoretical modeling or due to better observational constraints, this may allow us to determine whether primordial magnetic fields are needed to delay the build-up of Lyman α photons or not. As demonstrated above, comoving field strengths of the order 1 nG can delay the build-up of a Lyman α background by $\Delta z \sim 10$, which is significantly stronger than other mechanisms that might delay the formation of luminous objects. Lyman Werner feedback is essentially self-regulated and never leads to a significant suppression of star formation [Johnson et al., 2007b,a], X-ray feedback from miniquasars is strongly constrained from the observed soft X-ray background [Salvaterra et al., 2005], and significant heating from dark matter decay or annihilation would be accompanied by a significant amount of secondary ionization, resulting in a too large optical depth. On the other hand, an important issue is the question regarding the first sources of light. As shown in Schleicher et al. [2008b], massive Pop. III stars are needed to provide the correct reionization optical depth. On the other hand, an additional population of less massive stars with an IMF according to [Scalo, 1998; Kroupa, 2002; Chabrier, 2003] might be present. Such a population would emit more photons between the Lyman α and β line per stellar baryon [Leitherer et al., 1999], and could thus shift the build-up of a Lyman α background to an earlier epoch. The onset of efficient coupling via the Wouthuysen-Field effect thus translates into a combined constraint on the stellar population and the strength of primordial magnetic fields.

A further effect occurs at high redshifts, where the additional heat input from magnetic fields due to ambipolar diffusion and the decay of MHD turbulence increases the 21 cm signal. As gas decouples earlier from the radiation field, the difference between gas and radiation temperature is larger and collisions are more effective in coupling the spin temperature to the gas temperature. The determination of the 21 cm signal from this epoch is certainly challenging, as the foreground emission corresponds to temperatures which are higher than the expected 21 cm brightness temperature by several orders of magnitude. However, as pointed out in other works [Di Matteo et al., 2002; Oh & Mack, 2003; Zaldarriaga et al., 2004; Sethi, 2005; Shchekinov & Vasiliev, 2007], the foreground emission is expected to be featureless in frequency, which may allow for a sufficiently accurate subtraction. In this context, it may also to help to focus on the frequency gradient of the mean brightness temperature, rather than the 21 cm brightness temperature itself. Upcoming long-wavelength

3. INFLUENCE OF PRIMORDIAL MAGNETIC FIELDS ON 21 CM EMISSION[1]

experiments such as LOFAR, 21CMA (former PAST)[1], MWA[2], LWA[3] and SKA[4] may thus detect the additional heat from primordial magnetic fields in the neutral gas, or otherwise set new upper limits on primordial magnetic fields, perhaps down to $B_0 \sim 0.1$ nG. Like the 21 cm transition of hydrogen, rotational and ro-vibrational transitions of primordial molecules may create interesting signatures in the CMB as well [Schleicher et al., 2008c], which may provide a further test of the thermal evolution during the dark ages.

---

[1] http://web.phys.cmu.edu/past
[2] http://web.haystack.mit.edu/arrays/MWA/index.html
[3] http://lwa.unm.edu
[4] http://www.skatelescope.org

4

The chemistry of the early universe and its signatures in the CMB[1]

In this chapter[1], I explore a further possibility to observe signatures from the early universe. Similar as the 21 cm line of atomic hydrogen, molecules in the primordial gas can lead to some net absorption in the cosmic microwave background, which would lead to small deviations from a blackbody spectrum. Further, scattering of CMB photons with molecules can lead to a frequency-dependent change of the CMB power spectrum. These possibilities have been considered first by Maoli et al. [1994]. It has been proposed that H^- could lead to significant net absorption in the CMB, which would be detectable with present technology [Black, 2006]. Here, I calculate these effects based on a self-consistent derivation of the chemical abundances, and including the processes of spontaneous and stimulated emission. This more detailed model shows that no observational effects can be expected for satellites like Planck, as the absorption is strongly balanced by corresponding emission processes.

4.1 Introduction

The cosmic microwave background (CMB) is one of the most powerful tools of high-precision cosmology, as it allows one to determine the cosmological parameters, the power spectrum of initial fluctuations and various other quantities. It is thus important to have a detailed theoretical understanding of all effects that have a potential influence on CMB measurements. Following the WMAP 3 year and 5 year results [Spergel et al., 2007; Hinshaw et al., 2007;

---

[1]This chapter is based on an article by Schleicher et al., A&A, 490, 521, 2008, reproduced with permission © ESO.

4. THE CHEMISTRY OF THE EARLY UNIVERSE AND ITS SIGNATURES IN THE CMB[1]

Komatsu et al., 2008] that confirmed our standard picture of cosmology, we are looking forward to the precise measurement that will be performed with Planck[1] in only a few years. The measurement of the electron scattering optical depth allows one to constrain the effective reionization redshift and yields indirect information about an epoch that cannot yet be observed. Recalling that the cross sections of bound electrons can be larger by orders of magnitude compared to the cross section of free electrons, the optical depth due to molecules may provide information on the dark ages of the universe, in spite of the small molecular abundances. In fact, the early work of Maoli et al. [1994] suggested that the molecular opacities could smear out CMB fluctuations on the scale of the horizon, and at the same time create new secondary fluctuations due to the interaction with the velocity fields which are present in proto-clouds in the dark ages. This work had the intention to explain why no CMB fluctuations had been observed at that time.

Since then, there has been considerable progress both in the understanding of chemical processes in the early universe and the molecular abundances, as well as in the interaction of molecules with the CMB and the generation of spectral-spatial fluctuations. While recombination was originally examined by Peebles [1968] and Zeldovich et al. [1969], and improved in several follow-up works [Matsuda et al., 1971; Jones & Wyse, 1985; Sasaki & Takahara, 1993] based on analytic methods, today's computers allow a detailed treatment of the recombination process based on a reaction network that takes into account hundreds of energy levels for H, He and He^+, as in the work of Seager, Sasselov and Scott [Seager et al., 2000]. A simplified code reproducing the results of this detailed calculation is given by Seager, Sasselov and Scott [Seager et al., 1999]. In a recent series of papers, Switzer & Hirata [2008a]; Hirata & Switzer [2008]; Switzer & Hirata [2008b] considered the recombination of helium in great detail. Deviations of the CMB spectrum from a pure blackbody have also been considered in various works. Dubrovich [1975] considered the effect of hydrogen recombination lines. Rubiño-Martín et al. [2006, 2008] examined distortions due to helium and hydrogen lines in more detail, and Chluba & Sunyaev [2006] examined distortions due to the two-photon process. Wong et al. [2008] give a good overview of recent improvements and uncertainties regarding the recombination process.

Here, the main focus is on the postrecombination universe and possible imprints in the CMB from this period. The formation of H_2 during the dark ages has already been discussed by Saslaw & Zipoy [1967]. A more detailed treatment of molecules has been performed by Puy et al. [1993], Stancil et al. [1998] and by Galli & Palla [1998]. A useful collection of analytic formulae for estimating the abundance of various molecules after recombination was given by Anninos et al. [1997]. Recently, this problem was re-examined by Puy & Signore

---

[1] http://www.rssd.esa.int/index.php?project=planck

4.1 Introduction

[2007], and Hirata & Padmanabhan [2006] examined H_2 formation in more detail, taking into account the effects due to non-thermal photons.

Regarding the interaction of molecules with the CMB, the effects of various molecules due to their optical depths have been considered by Dubrovich [1994]. The enhancement of spectral-spatial fluctuations due to the luminescence effect, which is well-known from stars in reflection nebulae, has also been discussed by Dubrovich [1997]. Observational prospects for Herschel and ODIN have been discussed by Maoli et al. [2005], and the relevance for Planck has been assessed by Dubrovich et al. [2007]. The effects discussed were based on smear-out of primary CMB fluctuations due to the molecular optical depth and the generation of secondary anisotropies due to scattering with proto-objects in the universe. Mayer & Duschl [2005] provided a recent overview of different processes that may contribute to the opacity in primordial gas, and derived Rosseland and Planck mean opacities.

Black [2006] recently considered the influence of the bound-free transition of the negative hydrogen ion on the CMB, and found an optical depth of more than 10^{-5} at 10 cm^{-1}, which would have interesting implications for Planck and other CMB experiments. This contributed to our original motivation to examine this and other effects in more detail. As we will show below, however, the optical depth is due to H^- bound-free transitions is much smaller than reported by Black [2006], and must in addition be corrected for stimulated and spontaneous emission. On the other hand, there are other effects from H^- and HeH^+ that are close to observational relevance.

From a numerical point of view, the pioneering work of Anninos et al. [1997] provided a flexible and easily extendible scheme that is still widely used in state-of-the art simulations of the early universe, and which is also adopted in the public version of the Enzo code [O'Shea et al., 2004; Bryan et al., 1995; Norman et al., 2007][1]. This scheme is extended here to account for effects during recombination and the evolution of primordial chemistry in the homogeneous universe. We further provide an extended overview of the essential processes that may influence the CMB, determine the contributions from the most relevant species, and discuss the possibilities for detection with the Planck satellite. In Chapter 4.2, we provide a general discussion of the potential effects of primordial chemistry on the CMB. In Chapter 4.3, we present the general picture regarding the formation of the first molecules in the universe and give some analytic estimates for the abundances. In Chapter 4.4, we present the chemical network for our calculation, which includes some new rates for the HeH^+ molecules that are given in the appendix. Section 4.5 explains the numerical algorithm for the chemical network, and Chapter 4.6 presents the model abundances as a function of

---
[1] http://lca.ucsd.edu/portal/software/enzo

4. THE CHEMISTRY OF THE EARLY UNIVERSE AND ITS SIGNATURES IN THE CMB[1]

redshift. In Chapter 4.7, we explain our treatment of the different species and discuss the observational implications. Further discussion and outlook is provided in Chapter 4.8.

4.2 Imprints from primordial molecules on the CMB

The interaction with the CMB is crucial to determine the evolution and abundance of primordial molecules, and conversely, this interaction may also leave various imprints on the CMB photons while they travel through the dark ages. The most obvious imprint on the CMB is probably a frequency-dependent change in the observed CMB spectrum $I(v)$ due to absorption by a species M. An upper limit on this effect is given by

$$I(v) = B(v)e^{-\tau_M(v)}, \tag{4.1}$$

where $B(v)$ denotes the unaffected CMB spectrum and the optical depth $\tau_M(v)$ of species M at an observed frequency v is given by an integration over redshift as

$$\begin{aligned}\tau_M(v) &= \int dl\, \sigma_M\left[v_0(1+z)\right] n_M \\ &= \frac{n_{H,0}c}{H_0} \int_0^{z_f} f_M(z)\sigma_M\left[v_0(1+z)\right] \frac{(1+z)^2}{\sqrt{\Omega_\Lambda + (1+z)^3 \Omega_m}} dz,\end{aligned} \tag{4.2}$$

where $\sigma_M(v)$ is the absorption cross section of the considered species as a function of frequency, n_M the number density of the species, $n_{H,0}$ the comoving hydrogen number density, c the speed of light, H_0 Hubble's constant, f_M the fractional abundance of the species relative to hydrogen, and z_f the redshift at which it starts to form efficiently. Such an optical depth can be provided by resonant line transitions from molecules with high dipole moments like HeH$^+$ and HD$^+$, free-free processes or photodestruction of species like H$^-$, He$^-$ or HeH$^+$, which have a relatively low photodissociation threshold. However, Eq. 4.1 gives only an upper limit because absorption can be balanced by inverse processes (spontaneous and stimulated emission). A better estimate can be obtained by introducing an excitation temperature T_{ex} defined by

$$\frac{n_u}{n_l} = \frac{g_u}{g_l}\exp\left(-\frac{E_u - E_l}{kT_{ex}}\right) \tag{4.3}$$

where n_u and n_l denote the population of the upper and the lower level, g_u and g_l are the corresponding statistical weights and k is Boltzmann's constant. The excitation temperature is essentially determined by the ratio of collisional and radiative de-excitations. For molecules with non-vanishing dipole moments, the de-excitation is dominated by radiative transitions, and the level populations are in equilibrium with the radiation temperature. The

4.2 Imprints from primordial molecules on the CMB

frequency-dependent change in the radiation temperature is given by

$$\Delta T_r \sim -(T_r - T_{\text{ex}}) \tau_M(\nu). \tag{4.4}$$

This expression indicates a major obstacle for the detection of primordial molecules: those species with high dipole moments and thus high cross sections have an excitation temperature which is very close to the temperature of the CMB, while species with low dipole moments have a very low optical depth. Molecular resonant line transitions are thus unlikely to lead to a non-negligible net change in the radiation temperature. For photodestruction processes, this is different because the destruction is regulated by the CMB temperature, while the inverse formation processes are governed by the gas temperature. Thus, photodestruction can lead to a net change in the number of CMB photons. The same is true for free-free processes, which emit a blackbody spectrum according to the gas temperature, but absorb the spectrum of the radiation field.

In addition, as discussed by Maoli et al. [1994], the optical depth from primordial molecules may smear out primary fluctuations in the CMB if the optical depth acts in a way that effectively scatters the CMB photons. This is possible even in a situation where stimulated and spontaneous emission balance the absorption of molecules, such that there is no net change in the number of photons. It is thus somewhat complementary to the effect discussed above. However, it must be noted that it requires spontaneous emission to be important, as stimulated emission does not change the direction of the photons, and does not provide a mechanism for scattering. We thus emphasize the importance to correct for stimulated emission. As shown by Basu et al. [2004] for small angular scales, such an optical depth (corrected for stimulated emission) then leads to a change in the power spectrum given by

$$\Delta C_l \sim -2\tau C_l, \tag{4.5}$$

where the C_l's are the usual expansion coefficients for the observed power spectrum. As discussed by Maoli et al. [1994], such scattering processes can in addition generate secondary anisotropies which are proportional to the optical depth of the scattering processes. However, this effect is suppressed by more than three orders of magnitude, as it is also proportional to the ratio of the peculiar velocity to the speed of light. Dubrovich [1997] discussed the luminescence effect of various molecules which could potentially amplify the generation of secondary anisotropies. Based on the new abundances found in this work, we will give a basic estimate of this effect in Chapter 4.8. Basu [2007] considered in addition the effects of emission from molecules like HD and LiH^+, as HD in particular is an important coolant in cold primordial gas. Unfortunately, the effect seems to be negligible.

4. THE CHEMISTRY OF THE EARLY UNIVERSE AND ITS SIGNATURES IN THE CMB[1]

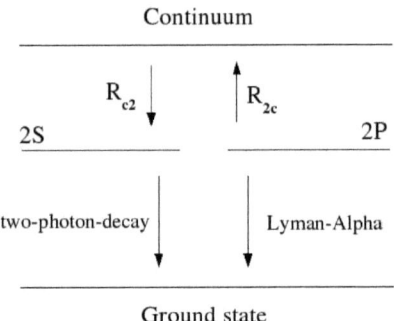

Figure 4.1: The idealized three-level hydrogen atom (ground state $1s$, excited states $2s$ and $2p$, continuum), and the relevant transitions.

4.3 Recombination and the formation of molecules in the early universe

4.3.1 Hydrogen recombination

Various approaches exist to calculate the time evolution of the ionized hydrogen fraction during recombination. The traditional approach is based on solving one single ordinary differential equation for the ionization degree x_e, which can be done using the approximation of an effective three-level-atom. This was first done by Peebles [1968], and an improved version of this equation was derived by Jones & Wyse [1985]. This treatment was based essentially on the following assumptions:

- It is sufficient to take into account only hydrogen (helium is not considered).
- Collisional processes are outweighted by radiative processes and may be ignored.
- The relative populations of the excited fine structure states are thermal.
- Recombination and photoionization rates to and from higher states are related by Saha's formula.
- Any such recombination cascades down to settle in the first excited state.
- The level populations obey $n_{1s} \ll n_{2s}$, with n_{1s} the number density of hydrogen atoms in the ground state and n_{2s} the corresponding density for the 2s state.

4.3 Recombination and the formation of molecules in the early universe

- Each net recombination gives rise to a Lyman-α photon or two others of lower energy.

Seager et al. [2000] presented a very detailed calculation independent of these a-priori assumptions, treating helium and hydrogen as multi-level atoms with several hundred levels and evolving one ordinary differential equation (ODE) for each level, using a self-consistent treatment of the radiation field and its interaction with matter, which effectively leads to a speed-up of recombination. To allow the integration of their new method in other cosmological applications like CMBFAST [Seljak & Zaldarriaga, 1996], Seager et al. [1999] produced a simplified version of this code which is capable of reproducing the results of the full multilevel treatment. This simplified code, RECFAST[1], solves three ODEs for the ionized hydrogen fraction x_p, the ionized helium fraction x_{He} and the gas temperature T. These ODEs are parametrized so as to reproduce the speed-up in the recombination process. In this paper, we make use of the ODE solving for the ionized hydrogen fraction, which is given by

$$\frac{dx_p}{dz} = \frac{[x_e x_p n_H \alpha_H - \beta_H (1-x_p) e^{-h_p \nu_{H,2s}/kT}]}{H(z)(1+z)[1 + K_H(\Lambda_H + \beta_H) n_H (1-x_p)]} \times [1 + K_H \Lambda_H n_H (1-x_p)]. \tag{4.6}$$

In this equation, n_H is the number density of hydrogen atoms and ions, h_p Planck's constant and the parametrized case B recombination coefficient for atomic hydrogen α_H is given by

$$\alpha_H = F \times 10^{-13} \frac{at^b}{1 + ct^d} \text{ cm}^3 \text{ s}^{-1} \tag{4.7}$$

with $a = 4.309$, $b = -0.6166$, $c = 0.6703$, $d = 0.5300$ and $t = T/10^4$ K, which is a fit by Pequignot et al. [1991] to the coefficient of Hummer [1994]. This coefficient takes into account that direct recombination into the ground state does not lead to a net increase of neutral hydrogen atoms, since the photon emitted in the recombination process can ionize other hydrogen atoms in the neighbourhood.

The fudge factor $F = 1.14$ serves to speed up recombination and is determined from comparison with the multilevel-code. The photoionization coefficient β_H is calculated from the recombination coefficient as $\beta_H = \alpha_H (2\pi m_e kT/h_p^2)^{3/2} \exp(-h_p \nu_{H,2s}/kT)$. The wavelength $\lambda_{H,2p}$ corresponds to the Lyman-α transition from the $2p$ state to the $1s$ state of the hydrogen atom. The frequency for the two-photon transition between the states $2s$ and $1s$ is close to Lyman-α and is thus approximated by $\nu_{H,2s} = c/\lambda_{H,2p}$, where c is the speed of light (i.e., the same averaged wavelength is used). Finally, $\Lambda_H = 8.22458$ s^{-1} is the two-photon rate for the transition $2s$-$1s$ according to Goldman [1989], $H(z)$ is the Hubble factor and $K_H \equiv \lambda_{H,2p}^3/[8\pi H(z)]$ the cosmological redshifting of Lyman α photons.

---

[1] http://www.astro.ubc.ca/people/scott/recfast.html

4. THE CHEMISTRY OF THE EARLY UNIVERSE AND ITS SIGNATURES IN THE CMB[1]

4.3.2 H_2 chemistry

Due to the expansion of the universe, not all of the free electrons will recombine with protons. Instead, a freeze-out occurs, as recombination becomes less efficient at lower densities. The freeze-out abundance of free electrons was fitted by Peebles [1993] as

$$\frac{n_{H^+}}{n_H} \sim 1.2 \times 10^{-5} \frac{\Omega_0^{1/2}}{h\Omega_b} \qquad (4.8)$$

for cosmologies with total mass density parameter Ω_0, baryonic density parameter Ω_b and h is the Hubble constant in units of 100 km/s/Mpc. There is no contribution from helium to the free electron fraction, as helium recombines very early. The free electron fraction leads to the formation of molecules and ions like H^- and H_2^+. As discussed by Anninos & Norman [1996] and Anninos et al. [1997], their abundances can be calculated by assuming chemical equilibrium, as their formation and destruction timescales are much shorter than the Hubble time. This yields the approximate expressions

$$\frac{n_{H^-}}{n_H} \sim 2 \times 10^{-9} T^{0.88} \frac{n_{H^+}}{n_H}, \qquad (4.9)$$

$$\frac{n_{H_2^+}}{n_H} \sim 3 \times 10^{-14} T^{1.8} \frac{n_{H^+}}{n_H}. \qquad (4.10)$$

The matter temperature can be calculated by assuming that at $z > 200$ it is the same as the radiation temperature, owing to efficient coupling of the temperatures through the Compton scattering of CMB photons off the residual free electrons, while at $z < 200$, where this coupling becomes ineffective, it evolves as for a simple adiabatic expansion [Sunyaev & Zeldovich, 1972; Anninos & Norman, 1996]. So

$$T = T_0(1+z), \ z > 200, \quad T = \frac{T_0}{1+200}(1+z)^2, \ z < 200, \qquad (4.11)$$

with $T_0 = 2.726$ K is the CMB temperature at $z = 0$ [Fixsen & Mather, 2002]. At redshifts $z \geq 100$, H^- is efficiently photodissociated by the CMB and H_2 is mainly formed by the process $H_2^+ + H \rightarrow H_2 + H^+$. Assuming that H_2^+ is formed most efficiently at redshift $z_0 = 300$ without being photo-dissociated by the CMB and that the hydrogen mass fraction is given by $f_H = 0.76$, one obtains for the H_2 abundance [Anninos & Norman, 1996]

$$\frac{n_{H_2}}{n_H} \sim 2 \times 10^{-20} \frac{f_H \Omega_0^{3/2}}{h\Omega_b}(1+z_0)^{5.1}. \qquad (4.12)$$

4.3.3 Deuterium chemistry

The HD abundance is mainly determined by the deuteration of hydrogen molecules (i. e. $H_2 + D^+ \to HD + H^+$). It is thus crucial to have the correct abundance of D^+, which is essentially determined by charge exchange with hydrogen atoms and ions, i.e. the processes $D + H^+ \to D^+ + H$ and $D^+ + H \to D + H^+$. As will be shown below, D^+ is very close to chemical equilibrium, which yields the abundance

$$\frac{n_{D^+}}{n_D} \sim 1.2 \times 10^{-5} \exp(-43 \text{ K}/T) \frac{\Omega_0^{1/2}}{h\Omega_b}, \qquad (4.13)$$

when expression (4.8) is used. The abundance of neutral deuterium atoms n_D can be determined by assuming that deuterium is almost fully neutral, i. e. $n_D \sim \Omega_b f_D \rho_c / m_D$, where f_D is the total mass fraction of deuterium, m_D the mass of one deuterium atom, $\rho_c = 3H_0^2/8\pi G$ the critical density and G is Newton's constant. It is clear from expression (4.13) that the abundance of D^+ drops exponentially at low temperatures. Direct deuteration of molecular hydrogen can thus only occur at redshifts where the exponential term is still of order one, before the exponential fall-off becomes significant. We thus evaluate the relative abundance at redshift 90. For both the formation and the destruction process $HD + H^+ \to H_2 + D^+$, we estimate the rates with the simple expressions of Galli & Palla [1998]. Again, we emphasize that more detailed numerical calculations should use the rates given in the appendix. The abundance is then given at $z = 90$ as

$$\left(\frac{n_{HD}}{n_H}\right)_{z=90} \sim 1.1 \times 10^{-7} f_D \exp(421 \text{ K}/T_{z=90}) \frac{f_H \Omega_0^{3/2}}{h\Omega_b}. \qquad (4.14)$$

As there is no efficient destruction mechanism for HD at lower redshifts, the fractional abundance remains almost constant for $z < 90$.

4.3.4 HeH$^+$ chemistry

As we will show below in more detail, chemical equilibrium is also an excellent approximation to determine the abundance of HeH$^+$. For the rates presented in Galli & Palla [1998], the process of stimulated radiative association of H^+ and He dominates over the non-stimulated rate. With the new rates presented in appendix B, we find that both rates roughly coincide for gas temperatures greater than 10 K. Thus, for an analytic estimate, we approximate the combined formation rate through stimulated and non-stimulated radiative association by taking twice the rate for non-stimulated radiative association. The dominant destruction process at low redshifts is charge-exchange via $HeH^+ + H \to He + H_2^+$, which yields for chemical

4. THE CHEMISTRY OF THE EARLY UNIVERSE AND ITS SIGNATURES IN THE CMB[1]

equilibrium

$$n_{\text{HeH}^+} \sim 1.76 \times 10^{-10} n_{\text{He}}(n_{\text{H}^+}/n_{\text{H}}) \left(\frac{T}{300 \text{ K}}\right)^{-0.24} \exp\left(-\frac{T}{4000 \text{ K}}\right)$$
$$\sim 7.03 \times 10^{-16} n_{\text{H}} \frac{\Omega_0^{1/2}}{h\Omega_B} \left(\frac{T}{300 \text{ K}}\right)^{-0.24} \exp\left(-\frac{T}{4000 \text{ K}}\right). \qquad (4.15)$$

4.4 The chemical network

In our chemical network, we have included the formation paths of H_2 and HD. The complete list of rates is given in Table B, and some new rates that are relevant for HeH^+ are discussed in more detail in the appendix. H_2 can be formed by two main channels, via the reactions $H^- + H \rightarrow H_2 + e$ and $H_2^+ + H \rightarrow H_2 + H^+$. A very good compilation for the H_2 chemistry was given by Yoshida et al. [2006]. Our compilation for the H_2 formation rates is similar, but we do not include all of their three-body-processes, as they are not relevant in the low-density regime explored here. Also, we keep those modifications for low temperatures that were originally given by Abel et al. [1997]. The ionized fraction of hydrogen is not determined by solving rate equations, but from the RECFAST code of Seager et al. [1999]. The photodissociation rates for most of the hydrogen species are those of Galli & Palla [1998]. For the photodissociation of H_2^+, we use the rate from their standard model, which assumes that the the levels of the molecule are populated according to LTE. Since the H_2^+ level populations will be strongly coupled to the CMB at the redshifts at which H_2^+ photodissociation is significant, this assumption is more reasonable than using a rate that assumes that H_2^+ is completely in the ground state. However, a better understanding of this molecule would certainly be desirable. For the molecule HD^+, we estimate the photodissociation due to reactions 22 and 23 of Table A.1 as half of the corresponding reaction for H_2^+. This is in agreement with recent isotopic helium experiments [Pedersen, 2005; Buhr, 2007], which found a similar effect for dissociative excitation with helium isotopes.

Photodissociation of H_2 by the Solomon process [reaction 20 Stecher & Williams, 1967] is calculated following the procedure described in Glover & Jappsen [2007], with the assumption that the rotational and vibrational levels of H2 have their LTE level populations.

In a recent work of Capitelli et al. [Capitelli et al., 2007], it was shown that the reaction rate for the process $H_2 + e \rightarrow H^- + H$ is larger by several orders of magnitude than the rate given by Galli & Palla [1998]. This is because they include the vibrational levels of molecular hydrogen in their calculation, and find that the excited vibrational levels cannot be neglected for this process. The change in the order of magnitude, however, does not lead to a significant change in the results, as this rate is multiplied by the rather small densities of molecular hydrogen and free electrons.

The hydrogen and helium chemistry is almost completely decoupled, as can be seen from the small rates for the charge exchange reactions 50 and 51. Their main interaction is via the HeH$^+$ molecule. Owing to its relatively high contribution to the optical depth for the CMB, we have included this molecule in our chemical network and present some new rate coefficients for it in the appendix. The molecule HD gives a contribution to the cooling and is mainly formed through by the deuteration reactions 30 and 32, but there are also contributions from reactions 31, 43 and 44, involving HD$^+$, D$^-$ and H$^-$. For the formation of deuterated molecules, it is therefore important to determine the ionized fraction of deuterium, which is given through the charge exchange reactions 27 and 28. Our deuterium network is inspired by the compilation of Nakamura & Umemura [2002]. The main difference compared to the deuterium network of Galli & Palla [1998] is the detailed treatment of D$^-$. We have added reaction 44, estimating its rate from reaction 43, thus also considering the contribution of H$^-$. Also, reactions 27 and 28 are calculated from the revised rates of Savin [2002]. As the fit provided by Savin [2002] for the rate of reaction 27 becomes negative for $T < 2.5$ K, it is set to zero at these temperatures[1]. For temperatures larger than 200 K, the revised set of deuterium rates of Galli and Palla [Galli & Palla, 2002] are used. For lower temperatures, however, some of the new rates show unphysical divergences. In these cases, we use the rates from Galli & Palla [1998] when the temperature drops below 200 K.

4.5 The numerical algorithm

To determine the evolution of primordial molecules in the early universe, we have employed the chemical network of the Enzo code [O'Shea et al., 2004; Bryan et al., 1995; Norman et al., 2007] and extended the numerical approach developed by Anninos et al. [1997] for primordial chemistry to determine the chemical evolution of the homogeneous universe. The main issue was the calculation of the ionization fraction, which is determined by complex interactions between the CMB and the ground state as well as the excited states of atomic hydrogen, and which goes beyond typical applications of primordial chemistry. We thus included the RECFAST code of Seager et al. [1999] as a subroutine for this calculation. For the deuterium and helium species as well as the molecules, however, we use the first order backwards differencing (BDF) method developed by Anninos et al. [1997]. The chemical timestep is set to 1% of the hydrodynamical timestep. The latter is given by $\Delta t_{\text{hydro}} = \eta(\Delta x/c_s)$, where Δx is the cell size, c_s is the sound speed and η is a safety factor, here taken to be 0.5. This proved sufficient to resolve the relevant chemical timescales, and

---

[1] Note that this does not introduce a significant error, as owing to its exponential dependence on temperature, rate 27 is tiny at $T < 2.5$ K.

4. THE CHEMISTRY OF THE EARLY UNIVERSE AND ITS SIGNATURES IN THE CMB[1]

simulations performed with even smaller timesteps gave identical results. The rate equations for the species i are given in the form

$$\frac{dn_i}{dt} = -D_i n_i + C_i, \qquad (4.16)$$

where D_i and C_i are the destruction and creation coefficient for species i, respectively, which in general depend on the number densities of the other species and on the radiation field. Equation 4.16 is discretized and the right-hand-side is evaluated at the new timestep, yielding

$$\frac{n^{\text{new}} - n^{\text{old}}}{\Delta t} = -D_i^{\text{new}} n^{\text{new}} + C_i^{\text{new}}. \qquad (4.17)$$

This can be solved for n_i^{new}:

$$n_i^{\text{new}} = \frac{C^{\text{new}} \Delta t + n_i^{\text{old}}}{1 + D^{\text{new}} \Delta t}. \qquad (4.18)$$

The coefficients D_i^{new} and C_i^{new} are in general not known, but can be approximated using the species densities from the old timestep, and those species from the new timestep which have already been evaluated. Anninos et al. [1997] has argued that H^- and H_2^+ can even be evaluated assuming chemical equilibrium, since their reactions rates are much faster than those of the other species. Assuming some species j in chemical equilibrium essentially means that $\dot{n}_j = 0$, yielding

$$n_{j,eq} = \frac{C_j}{D_j}. \qquad (4.19)$$

In our chemical model, we have some additional species which have sufficiently fast reaction rates: HeH^+, D^- and HD^+. To check the validity of this assumption, we calculate the formation of molecules in two simulations: one using the non-equilibrium prescription (4.18) for all species, and one using the equilibrium description (4.19) for the species with fast reaction rates. The results are presented in Chapter 4.6 and confirm that the abundances of these species can be derived assuming chemical equilibrium.

The BDF method is not a fully implicit numerical scheme, as several of the destruction and creation mechanisms must be approximated using the species densities from the previous timestep. However, we found that it is stable when the ionized fraction is provided from RECFAST. For consistency checks, we have varied the chemical timestep and explicitly ensured mass conservation for hydrogen, helium and deuterium, yielding consistent results. Note that mass conservation, charge conservation and positivity must be ensured as described by Anninos et al. [1997] if the ionized fraction is not provided from an independent routine.

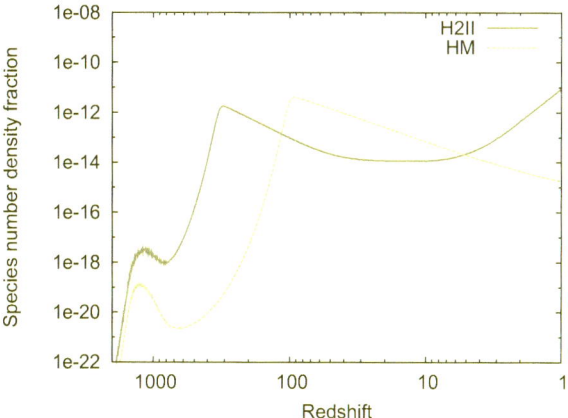

Figure 4.2: Results for the evolution of H_2^+ and H^-.

We summarize the numerical algorithm in the following way:

- Loop over the chemical timestep $t_{chem} = 0.01 t_{hydro}$ until the species have been evolved through the total hydrodynamical timestep.

- Update the temperature and the abundances of H, H^+ and e with RECFAST.

- Calculate the abundances of the atomic and ionized helium species using the non-equilibrium prescription (4.18), and the abundance of the molecule HeH^+ using (4.19).

- Calculate the abundances of H^- and H_2^+ using (4.19), and the abundance of H_2 using (4.18).

- Calculate D using (4.18), D^+, D^- and HD^+ using (4.19), and finally HD using (4.18).

4.6 Results from the molecular network

The formation of molecules after recombination was calculated for a ΛCDM model with $\Omega_{dm} = 0.222$, $\Omega_b = 0.044$, $\Omega_\Lambda = 0.734$, $H_0 = 70.9$ km/s/Mpc, $Y_p = 0.242$, [D/H] = 2.4×10^{-5}, where $\Omega_{dm}, \Omega_b, \Omega_\Lambda$ are the density parameters for dark and baryonic matter as

4. THE CHEMISTRY OF THE EARLY UNIVERSE AND ITS SIGNATURES IN THE CMB[1]

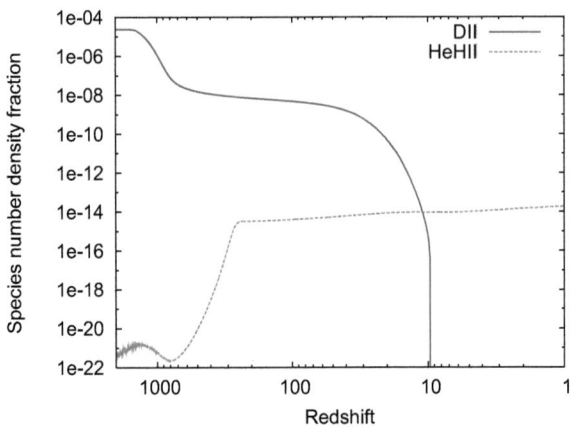

Figure 4.3: Results for the evolution of D^+ and HeH^+.

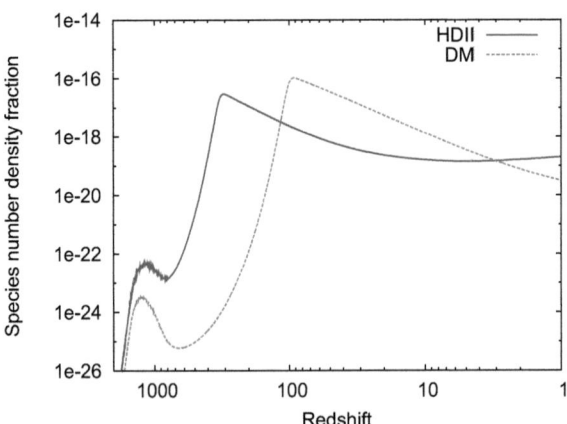

Figure 4.4: Results for the evolution of HD^+ and D^-.

4.6 Results from the molecular network

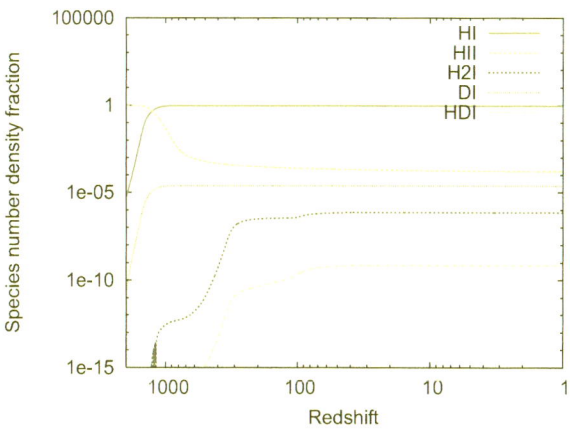

Figure 4.5: Results for those species which freeze-out and are not in chemical equilibrium.

Table 4.1: Freeze-out values of free electrons, H_2 and HD at z=100.

Cosmology / Ref.	[e /H]	[H_2/H]	[HD/H]
WMAP 3 + SDSS	2.63×10^{-4}	4.23×10^{-7}	2.39×10^{-10}
[Puy & Signore, 2007]	$\sim 10^{-4}$	1.5×10^{-7}	$\sim 2 \times 10^{-10}$
[Hirata, 2006]	-	$\sim 2 \times 10^{-8}$	-
Analytic approx.	2.0×10^{-4}	2.9×10^{-7}	1.6×10^{-10}
[Galli & Palla, 1998]	$\sim 2.4 \times 10^{-4}$	$\sim 1 \times 10^{-7}$	8×10^{-11}

Table 4.2: Freeze-out values of free electrons, H_2 and HD at z=10.

Cosmology / Ref.	[e /H]	[H_2/H]	[HD/H]
WMAP 3 + SDSS	1.92×10^{-4}	7.97×10^{-7}	7.52×10^{-10}
[Puy & Signore, 2007]	$\sim 6 \times 10^{-5}$	1.13×10^{-6}	3.67×10^{-10}
[Hirata, 2006]	-	$\sim 6 \times 10^{-7}$	-
[Seager et al., 1999]	1.92×10^{-4}	-	-
Analytic approx.	2.0×10^{-4}	2.9×10^{-7}	3.0×10^{-10}
[Galli & Palla, 1998]	3.02×10^{-4}	1.1×10^{-6}	1.21×10^{-9}

4. THE CHEMISTRY OF THE EARLY UNIVERSE AND ITS SIGNATURES IN THE CMB[1]

Table 4.3: Abundances of D^+ and HeH^+ at z=100.

Cosmology / Ref.	[D^+/H]	[HeH^+/H]
WMAP 3 + SDSS	5.15×10^{-9}	4.27×10^{-15}
[Puy & Signore, 2007]	$\sim 1.5 \times 10^{-9}$	$\sim 2.8 \times 10^{-15}$
Analytic approx.	3.7×10^{-9}	4.5×10^{-15}
[Galli & Palla, 1998]	$\sim 7.5 \times 10^{-9}$	$\sim 1.8 \times 10^{-14}$

Table 4.4: Abundances of D^+ and HeH^+ at z=10.

Cosmology / Ref.	[D^+/H]	[HeH^+/H]
WMAP 3 + SDSS	8.02×10^{-16}	9.98×10^{-15}
[Puy & Signore, 2007]	$\sim 1 \times 10^{-19}$	4.6×10^{-14}
Analytic approx.	2.1×10^{-20}	1.3×10^{-14}
[Galli & Palla, 1998]	$\sim 5 \times 10^{-17}$	$\sim 6.5 \times 10^{-14}$

well as dark energy, H_0 is Hubble's constant, Y_p is the mass fraction of helium with respect to the total baryonic mass and [D/H] is the mass fraction of deuterium relative to hydrogen. These are the parameters from the combined set of WMAP 3-year data and the data of the Sloan Digital Sky Survey (SDSS) [Spergel et al., 2007]. We expect only marginal changes in the results for the recently published WMAP 5-year data, as the cosmological parameters did not change significantly [Komatsu et al., 2008]. We have performed the calculation for the HD cooling functions of Galli & Palla [1998], Flower et al. [2000] and Lipovka et al. [2005], and find that the results are not sensitive to this choice. For our H_2 cooling function, we use that of Galli & Palla [1998], but H_2 cooling of the gas is never important, and our results would not change significantly if we were to use the revised cooling rate of Glover & Abel [2008]. The detailed evolution of the various species is plotted in Figs. 4.2-4.5. In Fig. 4.2, we give the results for H_2^+ and H^-, Fig. 4.3 shows the evolution of D^+ and HeH^+, and Fig. 4.4 the evolution of HD^+ and D^-. For the other species, the results are given in Fig. 4.5. For several interesting species, we give the results at $z = 100$ and $z = 10$ in tables 4.1-4.4, and compare them to the results of Galli & Palla [1998], Puy & Signore [2007], Seager et al. [1999] and the analytical approximations of Anninos & Norman [1996] and Chapter 4.3. As the analytic approximations do not take into account the H^- channel of H_2 formation, they underestimate the abundance found in the numerical calculation by roughly a factor of 2 at redshift 10.

The results clearly show the typical evolution of primordial chemistry as it is known from

4.6 Results from the molecular network

previous works. The two main channels for H_2 formation, via H^- and H_2^+, are reflected in its cosmic formation history, yielding a first major increase at $z \sim 300$, where the relative abundance of H_2^+ reaches a maximum, and a second major increase at $z \sim 100$, at a maximum of the H^- fraction. At redshifts below 5, there is a new rise in the abundance of H_2^+. This is likely an unphysical feature from the fit to the rate, which is not valid below 1 K. However, the real evolution at these low redshifts will in any case depart from our calculation due to reionization, metal enrichment and structure formation, and this feature is not relevant with respect to the CMB. The evolution of the deuterium species essentially follows the evolution of the hydrogen species. Since deuterium and hydrogen are strongly coupled via charge-exchange reactions, they recombine at almost the same time. However, due to the efficient charge-exchange reactions, there is no freeze-out of D^+. Instead, its abundance drops exponentially at low redshifts. D^- and HD^+ peak at the same redshifts as H^- and H_2^+, and the evolution of HD resembles closely the evolution of H_2^+, as the dominant HD formation channel is given by $H_2 + D^+ \rightarrow HD + H^+$. The evolution of HeH^+ consists of a first phase where its evolution is determined by the effectiveness of photodissociation, and a second phase where it is determined by charge-exchange with neutral hydrogen atoms. Due to the new formation rates presented in appendix B, formation through stimulated and non-stimulated radiative association of He and H^+ is almost equally important and in total less effective than found in previous works [Puy & Signore, 2007; Galli & Palla, 1998; Stancil et al., 1998].

Numerically, the free electron fraction found in this work roughly agrees with the analytic approximations of Anninos & Norman [1996] and previous results of Galli & Palla [1998], while Puy & Signore [2007] give a somewhat lower abundance. As the electrons act as catalysts for H_2 formation, this is reflected also in the abundance of molecular hydrogen, and similarly in the abundance of HD, as it is primarily formed by direct deuteration of molecular hydrogen. At $z = 100$, the H_2 abundance found by Hirata & Padmanabhan [2006] is still more than one order of magnitude below the abundance we find here, due to the effects of non-thermal photons, while at redshift 10, this effect is much less important and their result is only 25% below the value found here. The analytic estimate of Anninos et al. [1997] also underestimates the H_2 abundance at $z = 10$ by more than a factor of 2, as it does not take into account the H_2 formation via H^-. At $z = 100$, the abundances for D^+ are of the same order of magnitude, although some differences exist, which may be due to the differences in the rates for the charge-exchange reactions, as well as differences in the abundance of ionized hydrogen. For the HeH^+ molecule, the abundance is still comparable to previous results at redshift 100, but lower by a factor of 5 at $z = 10$.

We emphasize here that the abundance of H^- is similar to the one found by Black [2006]. The peaks in the abundance at $z \sim 100$ and $z \sim 1400$ are reproduced, and their height agrees with the results from our calculation when the physical number density given in his paper

4. THE CHEMISTRY OF THE EARLY UNIVERSE AND ITS SIGNATURES IN THE CMB[1]

is converted to the fractional abundance given in Fig. 4.2. We estimate the uncertainty from reading off the values of his Fig. 1 to be a factor 2-3, and we point out that this is not sufficient to explain the difference in the optical depth that we find below.

4.7 Effects of different species on the cosmic microwave background

4.7.1 Molecular lines

Due to the discreteness of the molecular lines, contributions to the optical depth arise only in narrow redshift intervals of the order $\Delta z/z \sim 10^{-5}$, corresponding to the ratio between the thermal linewidth to the frequency of the transition. Peculiar motions of the order $300 \times (1 + z)^{-1/2}$ km s^{-1} may further increase the effective linewidth by one or two orders of magnitude. This has no influence on the results presented here which depend only on the product of Δz with the profile function that scales with the inverse of the linewidth, but may induce additional anisotropies, as we discuss in Chapter 4.8. To evaluate Eq. 4.2 for a given frequency, we compute all the redshift intervals for which the photon frequency lies within an absorption line of the molecule, and add up the contributions from these frequencies. The relative importance of various molecules for the optical depth calculation can be estimated with the formulae of Dubrovich [1994]. The most promising candidate is the HeH$^+$ molecule, as it has a strong dipole moment and is formed from quite abundant species. Unfortunately, its destruction rate is very high as well. Another interesting molecule with a strong dipole moment is HD$^+$. However, as a deuterated molecule, it has an even lower abundance. H$_2$D$^+$ is not considered because it has an even lower abundance [Galli & Palla, 1998], and H$_3^+$ has both a low abundance and no dipole moment. The other molecule in our chemical network with a non-zero dipole moment is HD. However, its dipole moment is eight orders of magnitude smaller than that of HeH$^+$, and so even though its peak abundance is five orders of magnitude larger, its effects will still be negligible compared to those of HeH$^+$. We therefore do not consider it further. In spite of its strong dipole moment, LiH is also not considered, as it was already shown by Bougleux & Galli [1997] and Galli & Palla [1998] that its abundance is lower by roughly 10 orders of magnitude compared to the value assumed by Maoli et al. [1994], who discussed its potential relevance.

For HeH$^+$, we calculate the optical depth by using the large dataset provided by Engel et al. [2005], which allows one to derive the line cross sections from the Einstein coefficients. The line cross section $\sigma_{M,i}$ weighted by the level population for a transition from an initial state i with vibrational quantum number v'', rotational quantum number J'' and energy E'',

4.7 Effects of different species on the cosmic microwave background

to a final state f with vibrational quantum number v' and rotational quantum number J' is given by

$$\sigma_{M,i}(v) = \frac{1.3271 \times 10^{-12}(2J'+1)c^2}{Q_{vr}v^2}\exp\left(\frac{-E''}{kT_r}\right)$$
$$\times \left[1-\exp\left(-\frac{h_p v}{kT_r}\right)\right]A_{fi}\Phi(v-v_{fi}), \quad (4.20)$$

A_{fi} is the Einstein coefficient, v_{fi} is the frequency at line centre of the transition i, $\Phi(v-v_{fi}) = \frac{1}{\Delta v_D \sqrt{\pi}}\exp(-(v-v_{fi})^2/\Delta v_D^2)$ the profile function for the line width $\Delta v_D = \sqrt{2kT/m_M}v_{fi}/c$, m_M the mass of the molecule and Q_{vr} is the partition function, given by $Q_{vr} = \sum_i g_i \exp(-E_i/kT)$ with the degeneracies $g_i = 2J+1$. Again, we emphasize that the results are insensitive to the choice of the linewidth. The total cross section σ_M is obtained as a sum over all $\sigma_{M,i}$. For temperatures between 500 and 10000 K, we use the fit provided by Engel et al. [2005], while for lower temperatures, we do linearly interpolate between the values in their Table 5. The factor $\left[1-\exp\left(-h_p v/kT_r\right)\right]$ takes into account the correction for stimulated emission, which is especially relevant for the pure rotational transitions with low frequencies. As discussed in Chapter 4.2, this must be taken into account as stimulated emission does not change the direction of the emitted photons.

For HD$^+$, we use the same formalism as for HeH$^+$, but we determine the partition function from the accurate energy levels given by Karr & Hilico [2006]. Following Shu [1991], we use the transition moments $|D_{fi}|$ given by Colbourn & Bunker [1976] to determine the Einstein coefficients for the ro-vibrational lines as

$$A_{fi} = \frac{32\pi^3 v_{fi}^3}{3\hbar_p c^3}|D_{fi}|^2, \quad (4.21)$$

where $\hbar_p$ is the reduced Planck constant.

The Einstein coefficients for the pure rotational transitions are calculated with the dipole moment $D_0 = 0.86$ Debye of Dubrovich [1994] from

$$A_{0,J\to 0,J-1} = \frac{32\pi^3 v_{fi}^3}{3\hbar_p c^3}D_0^2\frac{J}{2J+1}. \quad (4.22)$$

4. THE CHEMISTRY OF THE EARLY UNIVERSE AND ITS SIGNATURES IN THE CMB[1]

4.7.2 The negative hydrogen ion.

There are two effects associated with the negative hydrogen ion than can affect the optical depth seen by CMB photons: the bound-free process of photodetachment that has also been discussed by Black [2006], and free-free transitions that involve an intermediate state of excited H$^-$, i. e.

$$H + e + \gamma \rightarrow (H^-)^* \rightarrow H + e. \qquad (4.23)$$

While the importance of the free-free process is well-known for stellar atmospheres, there has been little work on this process in the low-temperature regime. As the fit formulae given by John [1988] and Gingerich [1961] diverge at low temperatures, we have updated previous work of Dalgarno & Lane [1966] to calculate the free-free absorption coefficient for the low temperature regime as described in appendix A, and we use the fit of John [1988] to the calculation of Bell & Berrington [1987] for temperatures higher than 2000 K, where it is accurate within 1%. For the bound-free process, we use the fit of John [1988] to the calculations of Wishart [1979]. The treatment regarding absorption, spontaneous and stimulated emission is based on the expressions of Ruden et al. [1990], but with the updated cross sections mentioned above. Apart from the usual optical depth due to absorption, we introduce also effective optical depths due to stimulated and spontaneous emission. The following expressions have to be evaluated for the redshift z, and the frequency dependence is suppressed for simplicity. Of course, we emphasize that the observed frequencies at $z = 0$ must be related correctly to the physical frequencies at higher redshifts. As usual, the contribution to absorption is given as

$$d\tau_{bf,abs} = +n_{H^-}\sigma_{bf} ds, \qquad (4.24)$$

where ds is the cosmological line element, σ_{bf} the cross section for the bound-free transition and n_{H^-} the number density of H$^-$. We further introduce the Planck spectrum $B_T(\nu)$ of temperature T and frequency ν. The contribution to the effective optical depth from stimulated emission is then given as

$$d\tau_{bf,stim} = -(n_{H^-})_{LTE}\sigma_{bf} e^{-h_p\nu/kT} ds, \qquad (4.25)$$

where the LTE abundance $(n_{H^-})_{LTE}$ of H$^-$ is given as

$$(n_{H^-})_{LTE} = n_e n_H \frac{\lambda_e^3}{4} e^{h_p\nu_0/kT}, \qquad (4.26)$$

where $\lambda_e = \frac{h_p}{\sqrt{2\pi m_e kT}}$ is the thermal de Broglie wavelength and $\nu_0 = 0.754$ eV the binding energy of H$^-$. Note that the LTE abundance must be used here, as the processes of spontaneous and stimulated emission depend on the actual density of electrons and hydrogen atoms, and in general the correction for stimulated emission cannot be included as a factor

4.7 Effects of different species on the cosmic microwave background

of $(1 - \exp(-h\nu/kT))$ unless the H$^-$ ion has its LTE abundance [Ruden et al., 1990]. The effective optical depth due to spontaneous emission is further given as

$$d\tau_{bf,spon} = -(n_{H^-})_{LTE}\frac{2h_p\nu^3}{c^2 B_{T_r}(\nu)}\sigma_{bf}e^{-h_p\nu/kT}ds. \tag{4.27}$$

For the free-free effect, stimulated emission is already included in the rate coefficients $a_\nu(T)$ given by John [1988], which are normalized to the number density of neutral hydrogen atoms and the electron pressure. The effective contributions to the optical depth from absorption and emission are then given as

$$d\tau_{ff,abs} = +n_H n_e kT a_\nu(T) ds, \tag{4.28}$$

$$d\tau_{ff,em} = -n_H n_e kT a_\nu(T) ds \frac{B_{T_g}(\nu)}{B_{T_r}(\nu)}. \tag{4.29}$$

The absorption by free-free transitions is thus proportional to a black-body spectrum for the radiation temperature, while the emission produces a spectrum determined by the gas temperature. In the following, we will refer further to the optical depth from absorption, which we define as

$$\tau_{abs} = \int \left(d\tau_{bf,abs} + d\tau_{ff,abs}\right), \tag{4.30}$$

and the effective optical depth

$$\tau_{eff} = \int \left(d\tau_{bf,abs} + d\tau_{bf,stim} + d\tau_{bf,spon} + d\tau_{ff,abs} + d\tau_{ff,em}\right), \tag{4.31}$$

which includes the corrections for emission. While the optical depth due to absorption, τ_{abs}, is essentially responsible for photon scattering and a change in the power spectrum according to Eq. (4.5), the effective optical depth τ_{eff} leads to a net change in the radiation flux. The resulting change in the CMB temperature can be obtained by a linear expansion as

$$\Delta T = \tau_{eff}(\nu)\frac{B_{T_r}(\nu)}{(\partial B_{T_r}(\nu)/\partial T)_{T_r}} = f(\nu)\tau_{eff}(\nu)T_r, \tag{4.32}$$

where we have introduced a frequency dependent correction factor $f(\nu)$, which can be evaluated to first order as

$$f(\nu) = (1 - \exp\left[-h_p\nu/kT\right])\frac{kT_r}{h\nu}. \tag{4.33}$$

4. THE CHEMISTRY OF THE EARLY UNIVERSE AND ITS SIGNATURES IN THE CMB[1]

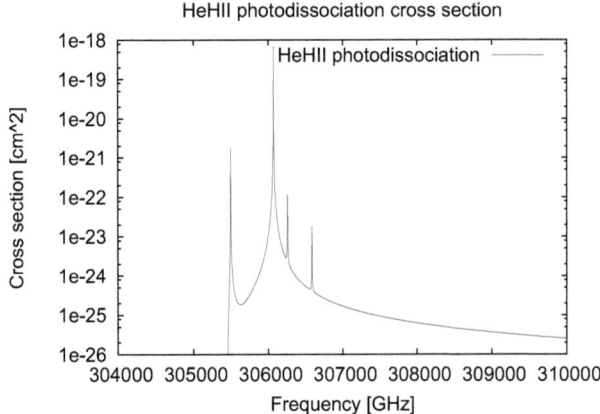

Figure 4.6: HeH$^+$ photodissociation cross section obtained from detailed balance.

4.7.3 The negative helium ion.

He$^-$ is to some degree similar to the H$^-$. However, only the free-free process contributes in practice, as any bound states autoionize on a timescale of the order of hundreds of microseconds [Holøien & Midtdal, 1955; Brage & Fischer, 1991]. Thus, we take into account only the free-free process, which can be treated in the same way as for H$^-$. We approximate the corresponding free-free coefficient by a power law proportional to v^{-2} and normalize with the data given by John [1994]. Such a treatment should be sufficient up to frequencies of 1000 GHz, and thus for the frequency range interesting for the Planck satellite.

4.7.4 Photodissociation of HeH$^+$.

We used detailed balance to determine the photodissociation cross section from inverse reaction. The latter is essentially determined by several narrow resonances, which have been tabulated by Zygelman et al. [1998]. The photodissociation cross section is thus given as the sum over resonances i as

$$\sigma_{ph}(v) = \sum_i \frac{m_e c^2 (h_p v - E_0)}{(h_p v)^2} \frac{\Gamma_{r,i} \Gamma_i / 2}{(h_p v - E_0 - E_{r,i})^2 + (\Gamma/2)^2}, \qquad (4.34)$$

4.7 Effects of different species on the cosmic microwave background

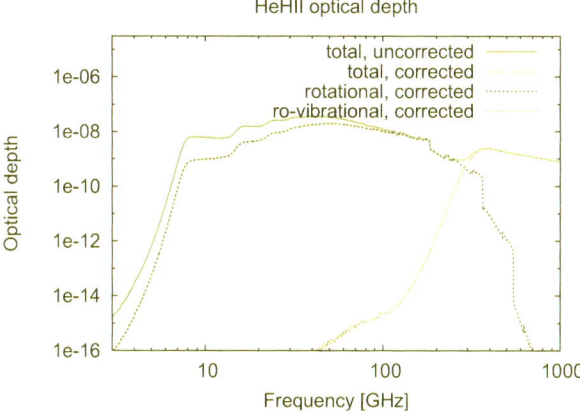

Figure 4.7: The HeH$^+$ optical depth, both corrected and uncorrected for stimulated emission. The contributions from pure rotational and ro-vibrational transitions are given separately.

where the parameters $E_{r,i}$, $\Gamma_{r,i}$ and Γ_i can be read off from Table 2 of Zygelman et al. [1998], and E_0 is the photodissociation threshold for HeH$^+$. From Dubrovich [1997], we adopt the value $E_0 = 1.85$ eV. The resulting photodissociation cross section is displayed in Fig. 4.6. We consider only absorption, as it is sufficient to rule out the contribution from this molecule with respect to Planck.

4.7.5 Observational relevance and results

As discussed in Chapter 4.2, the optical depth from resonant scattering can effect the CMB by a change in the power spectrum according to Eq. (4.5), and it may also produce secondary anisotropies. We neglect the latter effect for the moment, as it is suppressed by the ratio of the peculiar velocity over the speed of light. To quantify the importance of HeH$^+$ and HD$^+$ for a change in the power spectrum, we have calculated the optical depth and corrected for stimulated emission as described in the previous subsection. The results are given in Figs. 4.7 and 4.8. We find that the correction for stimulated emission is especially important for the lower frequencies of the pure rotational transitions. As explained by Basu et al. [2004], the sensitivity is not limited by cosmic variance when power spectra at different frequencies are compared, but the limit from instrumental noise corresponds to optical depths of 10^{-5} for the high-frequency bins. Thus, the signal is likely below the sensitivity of the Planck

4. THE CHEMISTRY OF THE EARLY UNIVERSE AND ITS SIGNATURES IN THE CMB[1]

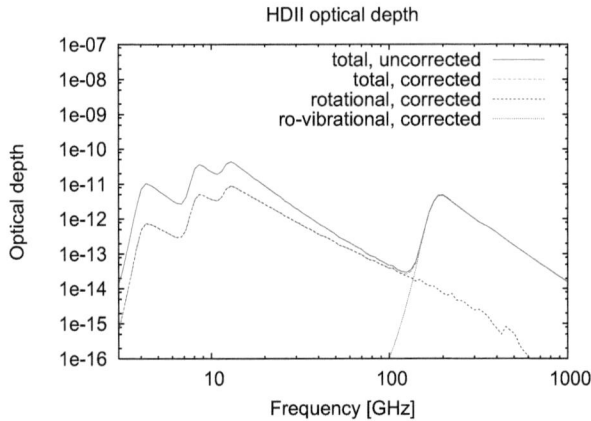

Figure 4.8: The HD$^+$ optical depth, both corrected and uncorrected for stimulated emission. The contributions from pure rotational and ro-vibrational transitions are given separately.

satellite by two orders of magnitude, but reasonable upper limits on the abundance of HeH$^+$ are feasible. From Figs. B.1.a and B.2.a in the appendix, we estimate the uncertainty in the formation rate to be a factor of 2, while the destruction rate might be larger by up to an order of magnitude, if the old values of Roberge and Dalgarno [Roberge & Dalgarno, 1982] are adopted. This defines the main uncertainty in this result. Even with an improved instrumental sensitivity, very accurate foreground subtraction would be required and may create additional noise.

Fig. 4.9 shows the optical depth due to absorption by several free-free and photodestruction processes. We find that it is dominated at low frequencies by the free-free contributions of H$^-$, and at high frequencies by the bound-free process of H$^-$. The optical depth from the free-free processes of H$^-$ and He$^-$ is essentially proportional to ν^{-2}, at least for frequencies smaller than 1000 GHz. This is essentially due to the characteristic frequency dependence in the absorption coefficient. Due to the approximation used for the He$^-$ free-free absorption coefficient, the corresponding optical depth is somewhat overestimated for larger frequencies, but it is still only a subdominant contribution to the total optical depth. The effect from the bound-free transition dominates, but is significantly lower than previously reported by Black [2006], and the reason for this discrepancy is not obvious: as pointed out in Chapter 4.6, our

4.7 Effects of different species on the cosmic microwave background

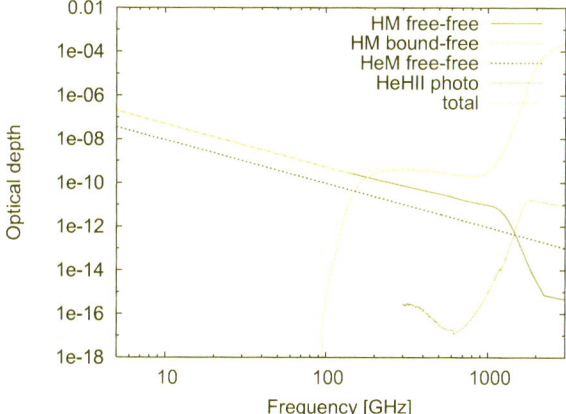

Figure 4.9: The absorption optical depth due to different processes: the free-free processes of H^- and He^-, the bound-free process of H^-, and the photodissociation of HeH^+. Clearly, the total optical depth is dominated by the processes involving H^-.

4. THE CHEMISTRY OF THE EARLY UNIVERSE AND ITS SIGNATURES IN THE CMB[1]

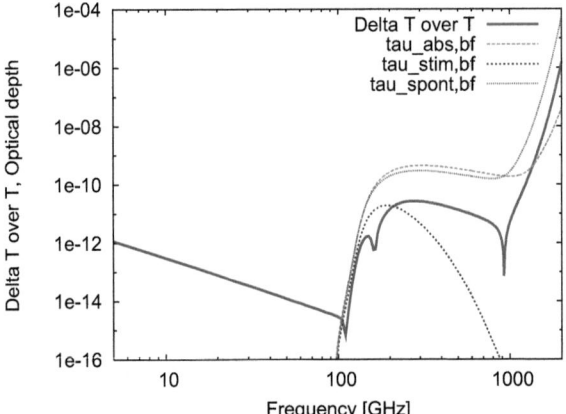

Figure 4.10: The relative change in the CMB temperature due to the presence of H$^-$. We further plot the optical depths due to absorption, spontaneous and stimulated emission for the bound-free process, as their overlap explains some features in the temperature change.

H$^-$ abundance agrees with his within a factor of 2 or 3, whereas the optical depth at high frequencies is different by three orders of magnitude. Although we have been unable to identify the reason for this disagreement, we have carefully checked our result and are confident that it is correct.

As discussed in Chapter 4.2, photodestruction and free-free processes can change the net number of CMB photons. The calculated change in the CMB temperature due to the free-free and bound-free effect is given in Fig. 4.10, and depends on the effective optical depth defined in Eq. (4.31). The change in the temperature due to the free-free transition is significantly lower than the absorption optical depth in Fig. 4.9, as the main contribution to the absorption optical depth comes from high redshifts $z > 300$, where the difference between radiation and gas temperature is very small and the optical depth due to absorption and emission balance each other. For the bound-free transition, the absorption is balanced to some extend by the spontaneous emission, and for frequencies larger than 1000 GHz, spontaneous emission in fact dominates, as the CMB flux at these frequencies is low by comparison. Near 150 and 1000 GHz, the contributions from spontaneous emission and absorption due to the bound-free process in fact become equal, making the net effect almost zero in a small frequency range.

The net change in the CMB temperature due to the free-free and bound-free processes of H$^-$ is small because absorption and emission processes are close to equilibrium. The optical depth due to absorption is thus considerably higher than the effective optical depth defined in Eq. (4.31), and can lead to a change in the power spectrum according to Eq. (4.5). Finally, we note that at redshifts where reaction 2 of Table B.1 dominates the destruction of H$^-$, there is an uncertainty of up to an order of magnitude in our predicted H$^-$ abundance, owing to the uncertainty in the rate of reaction 2 discussed in detail in Glover et al. [2006]. However, this uncertainty does not appear to significantly affect the size of our predicted signal, as the dominant contribution comes from redshifts at which reaction 2 is unimportant.

4.8 Discussion and outlook

We have provided a detailed network for primordial chemistry and solved for the evolution in the homogeneous universe, and we examined the various ways in which primordial species can influence the CMB. The detailed calculation in the previous sections suggests that the H$^-$ ion is only one order of magnitude below the detection threshold and strong upper limits on its abundance seem feasible, even though an accurate subtraction of frequency-dependent

4. THE CHEMISTRY OF THE EARLY UNIVERSE AND ITS SIGNATURES IN THE CMB[1]

foregrounds will be required for this purpose. The relative deviations of the CMB from a pure blackbody have been constrained by Mather et al. [1994]; Fixsen et al. [1996] to less than 1.5×10^{-5}. Distortions are also expected due to the two-photon process during recombination [Chluba & Sunyaev, 2006] and the helium and hydrogen lines [Rubiño-Martín et al., 2006, 2008]. Finestructure transitions in heavy elements are also expect to produce some scattering after recombination [Basu et al., 2004]. An accurate measurement of distortions in the CMB and a precise measurement of the CMB power spectrum can thus improve our understanding of various processes during and after recombination if an accurate foreground subtraction is feasible. Given the tiny change in the CMB temperature found in this work, we further conclude that detecting the change in the power spectrum caused by scattering (see Eq. (4.5)) is the most promising way to obtain constraints on the chemistry of the dark ages by future CMB experiments.

So far, we have only taken into account effects arising from the homogeneous universe. One might argue that the molecular abundances could be very different if there were a protogalaxy at the redshift of resonance. In fact, the redshift intervals for which resonance occurs are very narrow, corresponding to some 100 pc. This might be further increased by peculiar motions and local turbulence. Núñez-López et al. [2006] demonstrated that HD in cold collapsed clouds can lead to a strong local fluctuation of the order 10^{-5}. Given the small volume fraction of such clouds, we neglect their impact in the present paper, although we will examine the effect of such inhomogeneous fluctuations in future work.

Dubrovich [1997] suggested luminescence as an additional effect that may amplify the signal and lead to a strong frequency dependence. This effect is well-known for stars in reflection nebulae. It occurs at redshifts $z = 300 - 100$, where the rotational lines lie in the extreme Rayleigh-Jeans wing of the CMB and the first vibrational line near its maximum. Inelastic scattering in local velocity fields might thus provide much stronger frequency-dependent fluctuations. It is convenient to estimate the effect with the formula given by Dubrovich [1997]:

$$\frac{\Delta T_r}{T_r} = \left(\frac{\Delta T_r}{T_r}\right)_0 \frac{f_m}{10^{-10}} \frac{V_p}{30 \text{ km/s}} \frac{\Omega_b}{0.1}, \qquad (4.35)$$

where f_m is the fractional abundance of the molecule, V_p the peculiar velocity which can be estimated as $V_p = V_p(0)/\sqrt{1+z}$, $V_p(0) = 600$ km/s, and the quantity $(\Delta T_r/T_r)_0$ can be conveniently read off from Fig. 2 of Dubrovich [1997]. For HeH$^+$, it yields a maximum effect of roughly $\Delta T_r/T_r \sim 10^{-11}$, and $\Delta T_r/T_r \sim 10^{-12}$ for HD$^+$, which are clearly below the sensitivity of the Planck satellite.

As discussed by Launay et al. [1991], H_2 is in general formed in excited states, but quickly decays into the ground state. Thus, H_2 formation produces additional photons that may lie within the CMB radiation and thus produce distortions to the blackbody radiation. Prelimary

4.8 Discussion and outlook

estimates based on the transition between the first excited vibrational state and the ground state indicate that the effect yields a relative change in the CMB temperature of the order 10^{-15}, and is thus negligible.

At the end of the dark ages, reionization will dramatically change the chemical evolution of the intergalactic gas, and produce large regions of ionized gas. In such regions, other formation channels for molecules could be relevant, like $He^+ + H \rightarrow HeH^+ + \gamma$. The reaction rate of this channel is four orders of magnitude larger than the radiative association of H^+ and He. On the other hand, the influence of destruction processes, such as dissociative recombination and photodissociation, will also be enhanced. The details of the evolution will also depend on the ratio of stellar to quasar sources and the details of the transition from Pop III to Pop II stars. A detailed analysis of this contribution is beyond the scope of this work, but clearly there is the possibility that this epoch could further increase the optical depths of HeH^+ and H^-. Other contributions may arise from heavy elements at low redshift. Basu et al. [2004], Hernández-Monteagudo et al. [2006] and Basu [2007] suggested that one could use the change in the power spectrum of the CMB to constrain the chemical evolution of the low redshift universe. In fact, during the formation of metals and in early star forming regions, additional effects may occur that leave an interesting imprint in the CMB. As described by Hernández-Monteagudo et al. [2007a], oxygen pumping may change the CMB temperature in metal enriched environments in a similar way as the Wouthuysen-Field effect that is well-known from 21 cm studies [Wouthuysen, 1952; Field, 1958], and the inhomogeneous distribution of metallicity in bubble-like structures may influence the CMB power spectrum as described by Hernández-Monteagudo et al. [2008]. Hernández-Monteagudo et al. [2007b] further studied the effect of resonant scattering during reionization and recombination. In addition, star-forming regions may perturb the primordial signal through dust and molecular emission, especially CO [Righi et al., 2008a,b]. Such a potentially rich phenomenology will of course require a very careful analysis and a clear assessment of the different frequency dependences of various effects once the required sensitivity is reached. In the mean time, the increasing sensitivity in instruments like Planck [Bersanelli et al., 1996], the South Pole Telescope[1] [Ruhl et al., 2004] and the Atacama Cosmology Telescope[2] [Fowler, 2004] will allow at least to set upper limits that may constrain theories involving the dark ages, reionization and recombination.

---

[1] http://pole.uchicago.edu/
[2] http://www.physics.princeton.edu/act/

4. THE CHEMISTRY OF THE EARLY UNIVERSE AND ITS SIGNATURES IN THE CMB[1]

5

Dark stars: Implications and constraints from cosmic reionization and extragalactic background radiation[1]

In this chapter[1], observational constraints on dark stars powered by dark matter annihilation rather than nuclear fusion are examined in more detail. In particular, it is shown that it is difficult to reconcile scenarios that predict $\sim 1000\ M_\odot$ stars or lifetimes of ~ 100 Myr with the observed reionization constraints. I further discuss how the dark matter densities which are enhanced by adiabatic contraction during the formation of the stars can contribute to the cosmic gamma-ray background and the atmospheric neutrino background.

5.1 Introduction

Growing astrophysical evidence suggests that dark matter in the universe is self-annihilating. X-ray observations from the center of our Galaxy find bright 511 keV emission which cannot be attributed to single sources [Jean et al., 2006; Weidenspointner et al., 2006], but can be well-described assuming dark matter annihilation [Boehm et al., 2004a]. Further observations indicate also an excess of GeV photons [de Boer et al., 2005], of microwave photons [Hooper et al., 2007], and of positrons [Cirelli et al., 2008]. A common feature of these observations is that the emission seems isotropic and not correlated to the Galactic disk. However, there is usually some discrepancy between the model predictions and the amount

---
[1]This chapter is based on an article that is reprinted with permission from Schleicher, Banerjee & Klessen, PRD, 79, 043510, 2009. Copyright (2009) by the American Physical Society.

5. DARK STARS: IMPLICATIONS AND CONSTRAINTS FROM COSMIC REIONIZATION AND EXTRAGALACTIC BACKGROUND RADIATION[1]

of observed radiation, which may be due to uncertainties in the dark matter distribution, astrophysical processes and uncertainties in the model for dark matter annihilation [de Boer, 2008].

It is well-known that weakly-interacting massive dark matter particles may provide a natural explanation of the observed dark matter abundance [Drees & Nojiri, 1993; Kolb & Turner, 1990]. Calculations by Ahn et al. [2005] indicated that the extragalactic gamma-ray background cannot be explained from astrophysical sources alone, but that also a contribution from dark matter annihilation is needed at energies between 1-20 GeV. It is currently unclear whether this is in fact the case or if a sufficient amount of non-thermal electrons in active galactic nuclei (AGN) is available to explain this background radiation [Inoue et al., 2008]. Future observations with the FERMI satellite [1] will shed more light on such questions and may even distinguish between such scenarios due to specific signatures in the anisotropic distribution of this radiation [Ando et al., 2007].

The first stars have been suggested to have high masses of the order $\sim 100\ M_\odot$, thus providing powerful ionizing sources in the early universe [Abel et al., 2002; Bromm & Larson, 2004]. The effect of dark matter annihilation on the first stars has been explored recently in different studies. Spolyar et al. [2008] showed that an equilibrium between cooling and energy deposition from dark matter annihilation can always be found during the collapse of the proto-stellar cloud. This has been explored further by Iocco [2008] and Freese et al. [2008c], who considered the effect of scattering between baryons and dark matter particles, increasing the dark matter abundance in the star. Iocco et al. [2008] considered dark star masses in the range $5 \leq M_* \leq 600\ M_\odot$ and calculated the evolution of the pre-main-sequence phase, finding that the dark star phase where the energy input from dark matter annihilation dominates may last up to 10^4 yr. Freese et al. [2008a] examined the formation process of the star in more detail, considering polytropic equilibria and additional mass accretion until the total Jeans mass of $\sim 800\ M_\odot$ is reached. They find that this process lasts for $\sim 5 \times 10^5$ yr. They suggest that dark stars are even more massive than what is typically assumed for the first stars, and may be the progenitors for the first supermassive black holes at high redshift. Iocco et al. [2008], Taoso et al. [2008] and Yoon et al. [2008] have calculated the stellar evolution for the case in which the dark matter density inside the star is enhanced by the capture of addition WIMPs via off-scattering from stellar baryons. Iocco et al. [2008] followed the stellar evolution until the end of He burning, Yoon et al. [2008] until the end of oxygen burning and Taoso et al. [2008] until the end of H burning. Yoon et al. [2008] also took the effects of rotation into account. The calculations found a potentially very long lifetime of dark stars and correspondingly a strong increase in the number of UV photons that may contribute to reionization. Dark stars in the Galactic center have been discussed by Scott et al. [2008b,a].

---

[1] http://www.nasa.gov/mission_pages/GLAST/science/index.html

Such models for the stellar population in the early universe imply that the first luminous sources produce much more ionizing photons, and reionization starts earlier than for a population of conventional Pop. III stars. In fact, we recently demonstrated that reionization based on massive Pop. III can well reproduce the observed reionization optical depth [Schleicher et al., 2008b]. Increasing the number of ionizing photons per stellar baryon may thus reionize the universe too early and produce a too large reionization optical depth. This can only be avoided by introducing a transition to a stellar population which produces less ionizing photons, such that the universe can recombine after the first reionization phase. We therefore consider a double-reionization scenario in order to re-obtain the required optical depth. We discuss such models in Chapter 5.2 and demonstrate that some models of dark stars require considerable fine-tuning in reionization models in order to be compatible with the reionization optical depth from the WMAP [1] 5-year data [Nolta et al., 2009; Komatsu et al., 2008] and to complete reionization at redshift $z \sim 6$ [Becker et al., 2001] in Chapter 5.3. In Chapter 5.4, we show how such scenarios can be tested via 21 cm measurements.

A further consequence of the formation of dark stars is the steepening of the density profiles in minihalos [Freese et al., 2008a; Iocco, 2008], thus increasing the dark matter clumping factor with respect to standard NFW models. In Chapter 5.5, we estimate the increase in the clumping factor during the formation of dark stars and compare the calculation with our expectation for conventional NFW profiles and heavy dark matter candidates. In Chapter 5.6, we perform similar calculations for the light dark matter scenario. Further discussion and outlook is provided in Chapter 5.7.

5.2 The models

As discussed in the introduction, various models have been suggested for dark stars. The main difference between these models comes from considering or neglecting scattering between dark matter particles and baryons. In addition, it is not fully clear how important a phase of dark matter capture via off-scattering from baryons actually is, depending on further assumptions on the dark matter reservoir. In the following, we will thus distinguish between main-sequence dominated models and capture-dominated models.

5.2.1 Main-sequence dominated models

After an initial phase of equilibrium between cooling and heating from dark matter annihilation [Spolyar et al., 2008; Iocco, 2008; Freese et al., 2008c], the dark star will contract further while the dark matter annihilates away and the heating rate thus decreases. This duration of this adiabatic contraction (AC) phase is currently controversial: While Iocco et al.

---

[1] http://lambda.gsfc.nasa.gov/

5. DARK STARS: IMPLICATIONS AND CONSTRAINTS FROM COSMIC REIONIZATION AND EXTRAGALACTIC BACKGROUND RADIATION[1]

[2008] find it to be in the range of $(2-20) \times 10^3$ yr, Freese et al. [2008a] require about 10^6 yr. However, with a surface temperature of ~ 6000 K, the stars are rather cold in this phase, and thus will not contribute significantly to reionization. The uncertainty in the duration of the AC phase is therefore not crucial in this context.

If the elastic scattering cross section as well as the dark matter density around the star are sufficiently large, the star will enter a phase which is dominated by the capture of further dark matter particles. Such a scenario will be discussed in more detail in the next subsection. Here, we assume that the elastic scattering cross section is either too small, or that the dark matter reservoir near the star is not sufficient to maintain the capture phase for long. Then, the star will enter the main-sequence phase (MS), in which the luminosity is generated by nuclear burning. Stars with ~ 1000 $M_\odot$ are very bright in this phase, and emit $\sim 4 \times 10^4$ hydrogen-ionizing photons per stellar baryon during their lifetime [Bromm et al., 2001; Schaerer, 2002]. We will refer to stars of such type, which have only a short or even no phase driven by dark matter capture, as MS-dominated models.

For the case of MS-dominated models, we will focus essentially on the very massive stars suggested by Freese et al. [2008a]. For stars in the typical Pop. III mass range, it has been shown elsewhere [e. g. Schleicher et al., 2008b] that they are consistent with reionization constraints. A star with ~ 800 $M_\odot$ forming in a dark matter halo of $\sim 10^6$ $M_\odot$ corresponds to a star formation efficiency of 1%, which we adopt for this case.

5.2.2 Capture-dominated models

For a non-zero spin-dependent scattering cross section between baryons and dark matter particles, stars can capture additional WIMPs which may increase the dark matter density inside the star. For a cross section of the order 5×10^{-39} cm^2 and an environmental dark matter density of $\sim 10^{10}$ GeV cm^{-3}, this contribution becomes significant and alters the stellar evolution during the main sequence phase. We will refer to such a scenario as a capture-dominated (CD) model. These phases have been studied in detail by Iocco et al. [2008], Taoso et al. [2008] and Yoon et al. [2008]. They found that the number of ionizing photons produced by such stars may be considerably increased with respect to high-mass stars without dark matter annihilation effects, which is mostly due to a longer lifetime. In particular for dark matter densities of $(1-5) \times 10^{10}$ GeV cm^{-3} , the number of produced ionizing photons may be increased by up to two orders magnitude, while it decreases rapidly for larger dark matter densities, and the number of ionizing photons per baryon even drops below the value for Pop. II stars at.threshold densities of 1×10^{12} GeV cm^{-3}. As Yoon et al. [2008] found only a weak dependence on stellar rotation, we will not explicitly distinguish between models with and without rotation in the follow.

For the calculation of reionization, we will focus on some representative models of Yoon et al. [2008] in the following. However, we point out that there are still significant uncertainties in these models, in particular the dark matter parameters and the lifetimes of the stars. The latter should be seen as upper limits, as they assume that a sufficient reservoir of dark matter is available in the stellar neighborhood to allow for ongoing dark matter capture. This may however be disrupted by dynamical processes. An apparent disagreement of dark stars in the early universe with our reionization model may thus indicate that the stellar lifetimes are indeed smaller due to such processes.

5.3 Reionization constraints

In this section, we briefly review our reionization model and discuss reionization histories for main-sequence and capture-dominated models. These calculations implicitly assume annihilation cross sections of the order 10^{-26} cm^2 and dark matter particle masses of the order 100 GeV, the values which are typically adopted in dark star models. In such models, dark matter annihilation does not contribute to cosmic reionization [Schleicher et al., 2008b]. The chemistry in the pre-ionization era is thus unchanged and well-described by previous works [Puy et al., 1993; Galli & Palla, 1998; Stancil et al., 1998; Schleicher et al., 2008c], such that the initial conditions for star formation are unchanged. Considering higher annihilation cross sections essentially yields an additional contribution to the reionization optical depth, which would sharpen the constraints given below.

5.3.1 General approach

Our calculation of reionization is based on the framework developed by Schleicher et al. [2008b], which we have implemented in the RECFAST code [1] [Seager et al., 1999, 2000]. We will review here only those ingredients which are most relevant for this work. During reionization, the IGM consists of a two-phase medium, i. e. a hot ionized phase and a rather cold and overally neutral phase. The relative size of these phases is determined from the volume-filling factor Q_{H^+} of the H$^+$ regions [Shapiro & Giroux, 1987; Haiman & Loeb, 1997b; Barkana & Loeb, 2001; Loeb & Barkana, 2001; Choudhury & Ferrara, 2005; Schneider et al., 2006] as a function of redshift, given by

$$\frac{dQ_{H^+}}{dz} = \frac{Q_{H^+} C(z) n_{e,H^+} \alpha_A}{H(z)(1+z)} + \frac{dn_{ph}/dz}{n_H}, \quad (5.1)$$

---
[1] http://www.astro.ubc.ca/people/scott/recfast.html

5. DARK STARS: IMPLICATIONS AND CONSTRAINTS FROM COSMIC REIONIZATION AND EXTRAGALACTIC BACKGROUND RADIATION[1]

where $C(z) = 27.466 \exp(-0.114z + 0.001328z^2)$ is the clumping factor [Mellema et al., 2006], n_{e,H^+} the number density of ionized hydrogen, α_A the case A recombination coefficient [Osterbrock, 1989], $H(z)$ the Hubble function, n_H the mean neutral hydrogen density in regions unaffected by UV feedback and dn_{ph}/dz the UV photon production rate. Our model consists of ordinary differential equations (ODEs) for the evolution of temperature T and ionized fraction x_i in the overall neutral medium. For the application considered here, the dominant contribution to the effective ionized fraction $x_{eff} = Q_{H^+} + (1 - Q_{H^+})x_i$ and the effective temperature $T_{eff} = 10^4$ K $Q_{H^+} + T(1 - Q_{H^+})$ comes indeed from the UV feedback of the stellar population, i. e. from the hot ionized phase. According to Gnedin & Hui [1998] and Gnedin [2000], we introduce the filtering mass scale as

$$M_F^{2/3} = \frac{3}{a}\int_0^a da' M_J^{2/3}(a')\left[1 - \left(\frac{a'}{a}\right)^{1/2}\right], \qquad (5.2)$$

where $a = (1+z)^{-1}$ is the scale factor and M_J the thermal Jeans mass, given as

$$M_J = 2M_\odot \left(\frac{c_s}{0.2 \text{ km/s}}\right)^3 \left(\frac{n}{10^3 \text{ cm}^{-3}}\right)^{-1/2}. \qquad (5.3)$$

Here, c_s is the sound speed evaluated at temperature T_{eff}, in order to take into account the backreaction of heating on structure formation. In this framework, the production of UV photons can be described as

$$\frac{dn_{ph}/dz}{n_H} \sim \xi \frac{df_{coll}}{dz}, \qquad (5.4)$$

where $\xi = A_{He}f_* f_{esc} N_{ion}$, with $A_{He} = 4/(4 - 3Y_p) = 1.22$, N_{ion} the number of ionizing photons per stellar baryon, f_* is the star formation efficiency and f_{esc} the escape fraction of UV photons from their host galaxies. The quantity f_{coll} denotes the fraction of dark matter collapsed into halos, and is given as

$$f_{coll} = \text{erfc}\left[\frac{\delta_c(z)}{\sqrt{2}\sigma(M_{min})}\right], \qquad (5.5)$$

where $M_{min} = \min(M_F, 10^5 M_\odot)$, $\delta_c = 1.69/D(z)$ is the linearized density threshold for collapse in the spherical top-hat model and $\sigma(M_{min})$ describes the power associated with the mass scale M_{min}.

A relevant question in this context is also the role of Lyman-Werner (LW) feedback, which may suppress the star formation rate in low-mass halos. The role of such feedback has been addressed using different approaches. For instance, Machacek et al. [2001], O'Shea & Norman [2008] and Wise & Abel [2007b] have addressed this question employing numerical

5.3 Reionization constraints

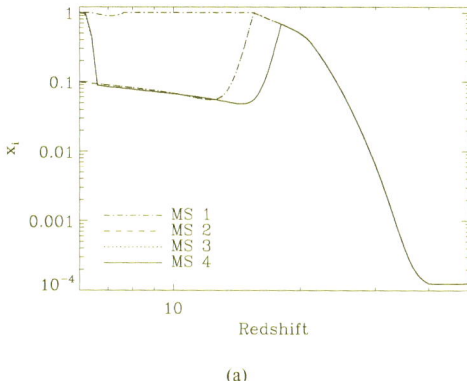

(a)

Figure 5.1: The evolution of the effective ionized fraction x_{eff}, for reionization models with main-sequence dominated dark stars (see Table 5.1). Models MS 1 and MS 2 can be ruled out by reionization constraints, while models MS 3 and MS 4 require a sudden increase in the star formation rate by a factor of 30 at redshift 6.5. It appears more realistic to assume lower masses and star formation efficiencies to reconcile dark star models with observations.

simulations in a cosmological context, assuming a constant LW-background radiation field. These simulations indicated that such feedback can delayed star formation considerably.

More self-consistent simulations show, however, that the above calculations overestimated the role of LW-feedback. Considering single stellar sources and neglecting self-shielding, Wise & Abel [2008] showed that LW-feedback only marginally delays star formation in halos that already started collapsing before the nearby star ignites. More detailed simulations taking into account self-shielding show that the star formation rate may be changed by only 20% in the presence of such feedback [Johnson et al., 2007b]. This is due to the rapid re-formation of molecular hydrogen in relic HII regions, which leads to abundances of the order 10^{-4}. Such abundances effectively shield against LW-feedback and make it ineffective [Johnson et al., 2007a]. This is the point of view adopted here, which may translate into an uncertainty of ~ 20% in the star formation rate. In fact, in scenarios involving dark matter annihilation, H_2 formation and self-shielding could be even further enhanced compared to the standard case [Mapelli & Ripamonti, 2007].

The models have to reproduce the reionization optical depth given by $\tau = 0.087 \pm 0.017$ [Komatsu et al., 2008] and fully ionization at $z \sim 6$ [Becker et al., 2001]. In the following, we will try to construct appropriate reionization histories for the different dark star models.

5. DARK STARS: IMPLICATIONS AND CONSTRAINTS FROM COSMIC REIONIZATION AND EXTRAGALACTIC BACKGROUND RADIATION[1]

5.3.2 Reionization with MS-dominated dark stars

As shown previously [Schleicher et al., 2008b], MS-dominated dark stars with $\sim 1000\ M_\odot$ would significantly overproduce the reionization optical depth if this type of stars had been common throughout the early universe. If, on the other hand, MS-dominated stars only had mass scales of $\sim 100\ M_\odot$, comparable to conventional Pop. III stars, reionization could not discriminate between them and conventional Pop. III stars, and dark stars would be compatible with observations. Alternatively, as explained in the introduction, a transition in the stellar population might help to alleviate the problem for high-mass dark stars. We will explore this possibility in more detail to work out whether such a scenario is conceiveable.

Numerical simulations by Dove et al. [2000], Ciardi et al. [2002] and Fujita et al. [2003] indicated rather high escape fractions of order 100% for massive Pop. III stars. Wood & Loeb [2000] found rather low escape fractions below 10%, while radiation hydrodynamics simulations by Whalen et al. [2004] show that such stars can easily photo-evaporate the minihalo. Here we adopt the point of view that indeed massive stars can photoevaporate small minihalos, but that the escape fraction will be reduced to $\sim 10\%$ in atomic cooling halos that have virial temperatures larger than 10^4 K. Thus, we set $f_{esc} = 1$ if the filtering mass is below the mass scale $M_c = 5 \times 10^7 M_\odot \left(\frac{10}{1+z}\right)^{3/2}$ that corresponds to the virial temperature of 10^4 K [Oh & Haiman, 2002; Greif et al., 2008], and $f_{esc} = 0.1$ in the other case. To reflect the expected stellar mass of $\sim 800\ M_\odot$, we choose a star formation efficiency of $f_* \sim 1\%$, an order of magnitude higher than what we expect for conventional Pop. III stars [Schleicher et al., 2008b].

Assuming that reionization is completely due to these MS-dominated dark stars (model MS 1), we find that the universe is fully ionized at redshift $z_{reion} = 15.5$ and the reionization optical depth is $\tau_{reion} \sim 0.22$, i. e. significantly larger than the WMAP 5 optical depth (see Fig. 5.1). Such a model is clearly ruled out.

To reconcile the presence of such massive dark stars with observations, one could invoke a double-reionization scenario, assuming a transition to a different mode of star formation induced by the strong UV feedback of MS-dominated dark stars. In fact, even for conventional star formation models, it is discussed that such UV feedback may lead to a less massive mode of star formation [Johnson & Bromm, 2006; Yoshida et al., 2007a,b]. In addition, chemical enrichment should facilitate such a transition as well [Schneider et al., 2006; Clark et al., 2008; Omukai et al., 2008; Smith et al., 2008a,b; Greif et al., 2009], although it is unclear how well metals will mix with the pristine gas. We assume that the transition to a low-mass star formation mode with a Scalo-type IMF [Scalo, 1998] happens at redshift 15.5, when the universe is fully ionized and UV feedback fully effective. For the subsequent Pop. II stars, we assume a star formation efficiency of $f_* = 5 \times 10^{-3}$ and $N_{ion} = 4 \times 10^3$ UV photons per stellar baryon.

5.3 Reionization constraints

Corresponding photon escape fractions are highly uncertain. Observations of Steidel et al. [2001] indicate an escape fraction of 10% at $z \sim 3$, while others find detections or upper limits in the range 5 – 10% [Giallongo et al., 2002; Malkan et al., 2003; Fernández-Soto et al., 2003; Inoue et al., 2005]. We adopt the generic value of 10% for simplicity, though our results do not strongly depend on this assumption. For this scenario, to which we refer as model MS 2, we find an optical depth $\tau_{\text{reion}} = 0.082$ well within the WMAP constraint, but the universe does not get fully ionized until redshift zero. This scenario is thus rejected based on the constraint from quasar absorption spectra [Becker et al., 2001].

To fulfill both the WMAP constraint as well as full-ionization at $z \sim 6$, we need to introduce an additional transition in our model. At redshift $z_{\text{burst}} = 6.5$, we increase the star formation efficiency to 15%. This might be considered as a sudden star burst and results in full-ionization at $z = 6.2$. In this case, we find $\tau_{\text{reion}} = 0.116$, which is within the 2σ range of the WMAP data. However, we are not aware of astrophysical models that provide a motivation for such a sudden star burst that increases the star formation rate by a factor of 30. Based on gamma-ray burst studies, Yüksel et al. [2008] showed that the cosmic star formation rate does not change abruptly in the redshift range between redshift zero and $z_{\text{burst}} = 6.5$. Such a sudden burst is thus at the edge of violating observation constraints.

To improve the agreement with WMAP, one can consider to shift the first transition to $z_{\text{Pop II}} = 18$ where full ionization is not yet reached (model MC 4), which yields an optical depth $\tau_{\text{reion}} = 0.086$, in good agreement with WMAP. At this redshift, 68% of the universe are already ionized, so UV feedback might already be active and induce a transition in the stellar population. The results are given in Fig. 5.1 and summarized in Table 5.1.

Model	$z_{\text{Pop II}}$	z_{burst}	τ_{reion}	z_f
MS 1	-	-	0.22	15.5
MS 2	15.8	-	0.078	never
MS 3	15.5	6.5	0.116	6.2
MS 4	18.	6.5	0.086	6.2

Table 5.1: Reionization models for MS-dominated dark stars. The parameters $z_{\text{Pop II}}$ and z_{burst} give the transition redshifts to a mode of Pop. II star formation and to the sudden star burst, while τ_{reion} is the calculated reionization optical depth and z_f the redshift of full ionization.

However, we find that only models MS 3 and MS 4 cannot be ruled out observationally. These models require two severe transitions in the stellar population and cannot be considered as "natural". Improved measurements of the reionization optical depth from Planck [1] will remove further uncertainties and may rule out model MS 3 as well. From a theoretical

---
[1] http://www.rssd.esa.int/index.php?project=planck

5. DARK STARS: IMPLICATIONS AND CONSTRAINTS FROM COSMIC REIONIZATION AND EXTRAGALACTIC BACKGROUND RADIATION[1]

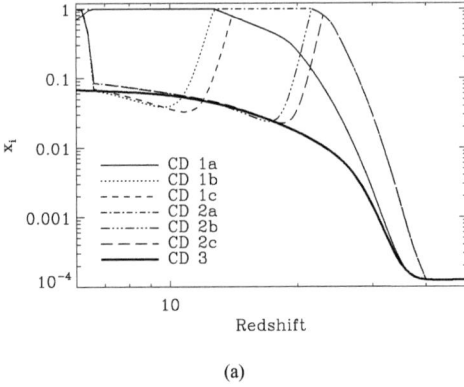

(a)

Figure 5.2: The evolution of the effective ionized fraction x_{eff}, for reionization models with capture-dominated dark stars (see Table 5.2). Models CD 1a, CD 1b, CD 2a, CD 2b and CD 3 are ruled out due to reionization constraints, while the remaining models require an artificial star burst.

point of view, it must be checked whether strong UV feedback can lead to the required transition to a low-mass star population, and in addition, the plausibility of a sudden star burst near redshift 6 must be examined as well. In summary, it seems more plausible to conclude that MS-dominated dark stars were less massive than suggested by Freese et al. [2008c], as already hinted by Schleicher et al. [2008b].

5.3.3 Reionization with CD dark stars

For CD dark star models, the situation is complicated by the fact that the number of UV photons per stellar baryon, N_{ion}, is model-dependent and changes with the environmental dark matter density, ρ_X. We select three representative models of Yoon et al. [2008], which assume a spin-dependent scattering cross section of 5×10^{-39} cm^2 (see Table 5.2). In general, stellar models depend on the product of this scattering cross section with the threshold dark matter density at the stellar radius [Taoso et al., 2008; Yoon et al., 2008]. Lower elastic scattering cross sections therefore correspond to going to smaller threshold densities at the same elastic scattering cross section.

In the models CD 1 and 2, N_{ion} is larger than for conventional Pop. III stars, while in the model CD 3, it is even less than in the case of Scalo-type Pop. II stars. Such a low luminosity

5.3 Reionization constraints

is unlikely to photo-evaporate star-forming halos, and we thus adopt $f_{esc} = 10\%$ for this case. However, such Scalo-type Pop. II stars are ruled out as sole sources for reionization [Schleicher et al., 2008b]. As we show in Fig. 5.2, even with a high star formation efficiency of $f_* = 1\%$, they never ionize the universe completely.

In principle, one could consider the presence of other sources to ionize the universe. While dark stars of type CD 3 may be the first stars to form, one might envision a transition to a stellar population with the power to ionize the universe. This transition is unlikely due to UV feedback, as UV feedback from dark stars is rather weak in this scenario. One thus has to rely on effective mixing of the produced metallicity, or assume that the first stellar clusters in atomic cooling halos contain a sufficient number of massive stars to reionize the universe [Clark et al., 2008].

For the other two models, N_{ion} is significantly larger and we adopt the procedure from the previous subsection, such that f_{esc} depends on the filtering mass. We adopt a star formation efficiency of $f_* = 0.1\%$. We examine the reionization models given in Table 5.2, which essentially follow the philosophy of the models from the previous section. We calculate the reionization history for the case where these dark stars are sole sources (CD 1a, CD2a) and find that the optical depth is considerably too high. We then determine the redshift where the universe is fully ionized and assume a transition to Pop. II stars at this redshift. In addition, to obtain full ionization at redshift 6, we assume a late star burst as in the models MS 3 and MS 4. This approach corresponds to the models CD 1b and CD 2b, and yields optical depth that are at least within the 2σ error of WMAP 5. In the models CD 1c and CD 2c, we improve the agreement with WMAP by introducing the Pop. II transition at an earlier redshift.

Reion. model	$\rho_X/10^{12}$	N_{ion}	f_*	$z_{\text{Pop II}}$	τ_{reion}
CD 1a	0.01 GeV cm^{-3}	1.75×10^5	0.1%	-	0.162
CD 1b	0.01 GeV cm^{-3}	1.75×10^5	0.1%	12.7	0.109
CD 1c	0.01 GeV cm^{-3}	1.75×10^5	0.1%	14.5	0.089
CD 2a	0.05 GeV cm^{-3}	2.4×10^6	0.1%	-	0.283
CD 2b	0.05 GeV cm^{-3}	2.4×10^6	0.1%	21.6	0.106
CD 2c	0.05 GeV cm^{-3}	2.4×10^6	0.1%	23	0.084
CD 3	1 GeV cm^{-3}	1.1×10^3	1%	-	0.004

Table 5.2: Reionization models for CD dark stars stars. The number of ionizing photons was determined from the work of Yoon et al. [2008]. The parameters $z_{\text{Pop II}}$ and z_{burst} give the transition redshifts to a mode of Pop. II star formation and to the sudden star burst, while τ_{reion} is the calculated reionization optical depth and z_f the redshift of full ionization. The calculation assumes a spin-dependent scattering cross section of 5×10^{-39} cm^2. As stellar models depend on the product of this cross section with the threshold dark matter density, the effect of a lower scattering cross section is equivalent to a smaller threshold density.

5. DARK STARS: IMPLICATIONS AND CONSTRAINTS FROM COSMIC REIONIZATION AND EXTRAGALACTIC BACKGROUND RADIATION[1]

The results are given in Fig. 5.2. Again, it turns out that somewhat artificial models are required to allow for an initial population of CD dark stars. The best way to reconcile these models with the constraints from reionization might be to focus on those models that predict a parameter $N_{\rm ion}$ which is closer to the Pop. III value of 4×10^4. This may be possible, as the transition from the models CD 1 and 2 to CD 3 is likely continuous, and an appropriate range of parameters may exist to re-concile models with observations. This would require a ρ_X between 10^{11} GeV cm^{-3} and 10^{12} GeV cm^{-3}. As mentioned earlier, the apparent violation of reionization constraints by some models depends also on the uncertainties in the stellar lifetime. If the dark matter reservoir near the star is destroyed earlier due to dynamical processes, the lifetime may be significantly reduced. Also, we stress that the conclusions depend on the adopted elastic scattering cross section and the dark matter density in the environment. The discussion here is limited to those models that have previously been worked out in detail.

5.4 Predictions for 21 cm observations

While some of the models suggested above essentially co-incide with standard reionization by mimicing the effects of conventional Pop. III stars, others may have a very distinctive signature, as they consist of a double-reionization phase, and upcoming 21 cm telescopes like LOFAR [1] or SKA [2] can thus verify or rule out such suggestions. The calculation shown in Fig. 5.3 is based on the double reionization model MS 4, but clearly the models MS 3, CD 1b, CD 1c, CD 2b and CD 2c yield similar results. In such a double-reionization scenario, the gas is heated to $\sim 10^4$ K during the first reionization epoch. Assuming that the first reionization epoch ends at redshift $z_{\rm Pop\ II}$, the gas temperature in the non-ionized medium will then evolve adiabatically as

$$T \sim 10^4 \text{ K} \left(\frac{1+z}{1+z_{\rm Pop\ II}} \right)^2. \tag{5.6}$$

In addition, the previous reionization phase will have established a radiation continuum between the Lyman α line and the Lyman limit, where the universe is optically thin, apart from single resonances corresponding to the Lyman series. This radiation is now redshifted into the Lyman series and may couple the spin temperature $T_{\rm spin}$ of atomic hydrogen to the gas temperature T via the Wouthuysen-Field effect [Wouthuysen, 1952; Field, 1958]. In fact, a small amount of Lyman α radiation suffices to set $T_{\rm spin} = T$ [Barkana & Loeb, 2005b; Hirata, 2006; Pritchard & Furlanetto, 2006], which we assume here. Also, as the universe

---
[1] http://www.lofar.org/
[2] http://www.skatelescope.org/

5.4 Predictions for 21 cm observations

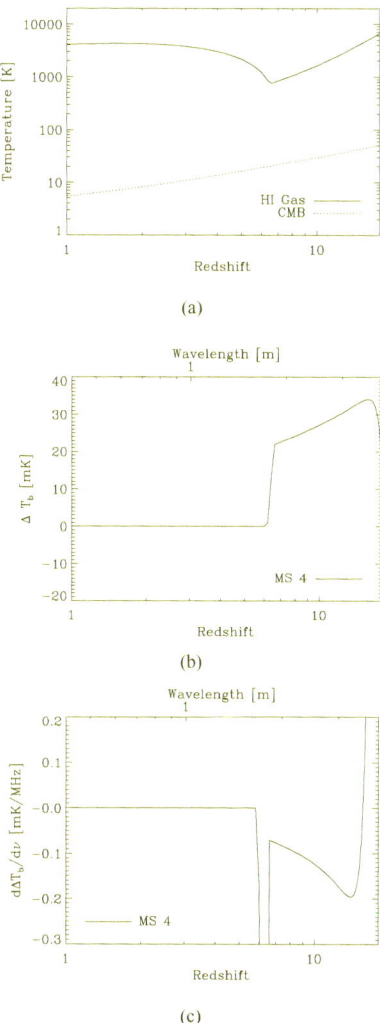

Figure 5.3: 21 cm signatures of double-reionization scenarios (here MS 4 from Table 5.1). Given is the evolution after the first reionization phase, when the H gas is heated from the previous ionization. Top: HI gas temperature, here identical to the spin temperature. Middle: Expected mean 21 cm brightness fluctuation. Bottom: Frequency gradient of the mean 21 cm brightness fluctuation.

5. DARK STARS: IMPLICATIONS AND CONSTRAINTS FROM COSMIC REIONIZATION AND EXTRAGALACTIC BACKGROUND RADIATION[1]

is optically thin to this radiation background, even Pop. II sources will suffice to couple the spin temperure to the gas temperature. The mean 21 cm brightness temperature fluctuation is then given as

$$\begin{aligned} \delta T_b &= 27 x_H (1+\delta) \left(\frac{\Omega_b h^2}{0.023}\right) \left(\frac{0.15}{\Omega_m h^2} \frac{1+z}{10}\right)^{1/2} \\ &\times \left(\frac{T_S - T_r}{T_S}\right) \left(\frac{H(z)/(1+z)}{dv_\parallel / dr_\parallel}\right) \text{ mK}, \end{aligned} \quad (5.7)$$

where x_H denotes the neutral hydrogen fraction, δ the fractional overdensity, Ω_b, Ω_m the cosmological density parameters for baryons and total matter, h is related to the Hubble constant H_0 via $h = H_0/(100 \text{km/s/Mpc})$, T_r the radiation temperature and $dv_\parallel / dr_\parallel$ the gradient of the proper velocity along the line of sight, including the Hubble expansion. We further calculate the frequency gradient of the mean 21 cm brightness temperature fluctuation to show its characteristic frequency dependence. In Fig. 5.3, we show the evolution of the gas temperature, the mean 21 cm brightness fluctuation and its frequency gradient for model MS 4.

As pointed out above, we expect similar results for other double-reionization models because of the characteristic adiabatic evolution of the gas and spin temperature. The decrease of the spin temperature with increasing redshift is a unique feature that is not present in other models that like dark matter decay [Furlanetto et al., 2006b] or ambipolar diffusion heating from primordial magnetic fields [Sethi & Subramanian, 2005; Tashiro et al., 2006; Schleicher et al., 2009a], which may also increase the temperature during and before reionization.

5.5 Cosmic constraints on massive dark matter candidates

In typical dark star models, it is assumed that massive dark matter candidates like neutralinos with masses of the order 100 GeV annihilate into gamma-rays, electron-positron pairs and neutrinos [Spolyar et al., 2008; Iocco, 2008; Freese et al., 2008b,a,c; Iocco et al., 2008; Yoon et al., 2008]. Similar to the constraint on high-redshift quasars from the X-ray background [Dijkstra et al., 2004; Salvaterra et al., 2005; Schleicher et al., 2008d], the gamma-ray and neutrino backgrounds allow to constrain the model for and the amount of dark matter annihilation. As detailed predictions for the decay spectra are highly model-dependent, it is typically assumed that roughly 1/3 of the energy goes into each annihilation channel. Constraints on such scenarios are available from the Galactic center and the extragalactic gamma-ray and neutrino backgrouns [Ullio et al., 2002; Beacom et al., 2007; Yüksel et al., 2007; Mack et al., 2008]. In this section, we consider how such constraints are affected when

5.5 Cosmic constraints on massive dark matter candidates

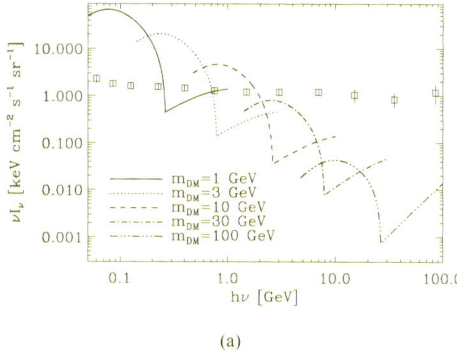

(a)

Figure 5.4: The predicted gamma-ray background due to direct annihilation into gamma-rays in the presence of adiabatic contraction during the formation of dark stars, and the background measured by EGRET (squares) [Strong et al., 2004]. One finds two peaks in the annihilation background for a given particle mass: One corresponding to annihilation at redshift zero, and one corresponding to the redshift where the enhancement from adiabatic contraction was strongest.

the increase in the annihilation rate due to enhanced dark matter densities after the formation of dark stars is taken into account.

5.5.1 Gamma-ray constraints

We adopt the formalism of Mack et al. [2008] who recently addressed the direct annihilation of massive dark matter particles into gamma-rays. The background intensity I_ν is given from an integration along the line of sight as

$$I_\nu = \frac{c}{4\pi} \int \frac{dz P_\nu([1+z]\nu, z)}{H(z)(1+z)^4}, \qquad (5.8)$$

where $P_\nu(\nu, z)$ is the (proper) volume emissivity of gamma-ray photons, which is given as

$$P_\nu = \alpha_b \delta((1+z)\nu - m_{\rm DM}) \frac{m_{\rm DM}}{\rm keV} \, {\rm keV} \, \langle \sigma v \rangle n_{\rm DM}^2 C_\gamma, \qquad (5.9)$$

where $\langle \sigma v \rangle = 3 \times 10^{-26}$ cm^3 s^{-1} denotes the thermally-averaged annihilation cross section, $\alpha_b = 1/3$ is the adopted branching-ratio to gamma-rays and $m_{\rm DM}$ the mass of the dark matter

5. DARK STARS: IMPLICATIONS AND CONSTRAINTS FROM COSMIC REIONIZATION AND EXTRAGALACTIC BACKGROUND RADIATION[1]

particle in keV. C_γ refers to the dark matter clumping factor. This clumping factor depends on the adopted dark matter profile and the assumptions regarding substructure in a halo [Ahn & Komatsu, 2005a; Ando, 2005; Chuzhoy, 2008; Cumberbatch et al., 2008]. Here we use the clumping factor for a NFW dark matter profile [Navarro et al., 1997] which has been derived by Ahn & Komatsu [2005b,a]. For $z < 20$, it is given in the absence of adiabatic contraction as a power-law of the form

$$C_{DM} = C_{DM}(0)(1+z)^{-\beta}, \qquad (5.10)$$

where $C_{DM}(0)$ is the clumping factor at redshift zero and β determines the slope. For a NFW profile [Navarro et al., 1997], $C_{DM}(0) \sim 10^5$ and $\beta \sim 1.8$. The enhancement due to adiabatic contraction is taken into account by defining

$$C_\gamma = C_{DM} f_{enh}, \qquad (5.11)$$

where the factor f_{enh} describes the enhancement of the halo clumping factor due to adiabatic contraction (AC). We have estimated this effect based on the results of Iocco et al. [2008], comparing a standard NFW profile with the enhanced profile that was created during dark star formation. We only compare them down to the radius of the dark star and find an enhancement of the order $\sim 10^3$. For the NFW case, the clumping factor would be essentially unchanged when including smaller radii as well, while the AC profile is significantly steeper and the contribution from inside would dominate the contribution to the halo clumping factor. However, as the annihilation products are trapped inside the star, it is natural to introduce an inner cut-off at the stellar radius. In addition, we have to consider the range of halo masses and redshifts in which dark stars may form. We assume that the halo mass must be larger than the filtering mass to form dark stars. However, there is also an upper mass limit. Halos with masses above

$$M_c = 5 \times 10^7 M_\odot \left(\frac{10}{1+z}\right)^{3/2} \qquad (5.12)$$

correspond to virial temperatures of 10^4 K [Oh & Haiman, 2002] and are highly turbulent [Greif et al., 2008]. It seems thus unlikely that stars will form on the very cusp of the dark matter distribution in such halos, and more complex structures may arise. We thus assume that dark stars form in the mass range between M_F and M_c. Once M_c becomes larger than M_F, dark star formation must end naturally. In fact, it may even end before, as discussed in Chapter 5.3. To obtain the highest possible effect, we assume that dark stars form as long as possible. We thus have

$$f_{enh} = \left(1 + 10^3 \frac{f_{coll}(M_F) - f_{coll}(M_c)}{f_{coll}(M_F)}\right). \qquad (5.13)$$

5.5 Cosmic constraints on massive dark matter candidates

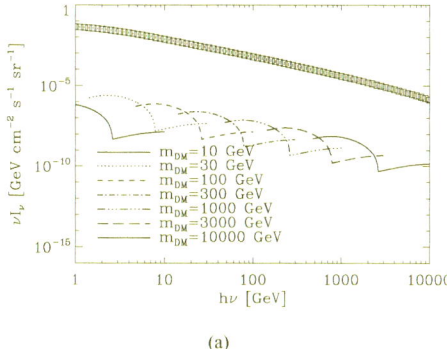

(a)

Figure 5.5: The predicted neutrino background due to direct annihilation into neutrinos in the presence of adiabatic contraction during the formation of dark stars, and the atmospheric neutrino background [Honda et al., 2004]. One finds two peaks in the annihilation background for a given particle mass: One corresponding to annihilation at redshift zero, and one corresponding to the redshift where the enhancement from adiabatic contraction was strongest.

In Fig. 5.4, we compare the results with EGRET observations of the gamma-ray background [Strong et al., 2004]. In the absence of adiabatic contraction, the predicted background peaks at the contribution from redshift zero [Mack et al., 2008]. We find that the enhancement of annihilation due to adiabatic contraction produces a second peak in the predicted background which originates from higher redshifts. In this scenario, particle masses smaller than 30 GeV can thus be ruled out.

5.5.2 Neutrino constraints

The contribution to the cosmic neutrino flux can be obtained in analogy to Eq. (5.8). As recent works [Beacom et al., 2007; Yüksel et al., 2007], we adopt an annihilation spectrum of the form

$$P_\nu = \alpha_b \delta((1+z)\nu - m_{\text{DM}}) \frac{m_{\text{DM}}}{\text{keV}} \text{ keV} \langle \sigma v \rangle n_{\text{DM}}^2 C_{\text{neutrino}}, \qquad (5.14)$$

which is analogous to the spectrum for annihilation into gamma-rays. The branching ratio to neutrinos is assumed to be 1/3 as well, and the annihilation comes from the same dark matter distribution, thus yielding $C_{\text{neutrino}} = C_\gamma$. The atmospheric neutrino background has been calculated from different experiments with generally good agreement [Ahrens et al., 2002; Gaisser & Honda, 2002; Honda et al., 2004; Ashie et al., 2005; Achterberg et al., 2007].

5. DARK STARS: IMPLICATIONS AND CONSTRAINTS FROM COSMIC REIONIZATION AND EXTRAGALACTIC BACKGROUND RADIATION[1]

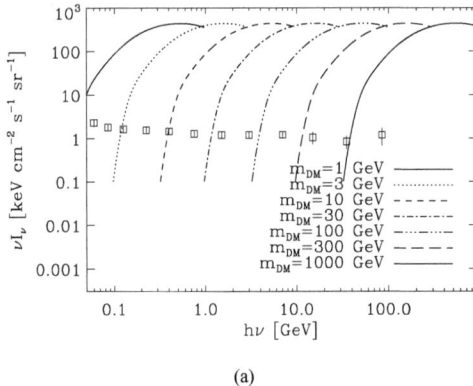

(a)

Figure 5.6: The *maximum* gamma-ray background due to direct annihilation into gamma-rays in the remnants of dark stars, and the background measured by EGRET [Strong et al., 2004]. The actual contribution to the gamma-ray background is highly model-dependent (see discussion in the text).

Iocco [2008] adopted a similar atmospheric neutrino flux for comparison with the expected neutrino flux from dark stars. We adopt here the data provided by Honda et al. [2004] and compare them to the predicted background in Fig. 5.5. The predicted background is always well below the observed background.

5.5.3 Emission from dark star remnants

In the previous subsections, we have included the enhancement of the halo clumping factor down to the stellar radius, as by definition the annihilation products on smaller scales are trapped inside the star. At the end of their lifetime, these stars may explode and the baryon density in the center may be largely depleted. The dark matter density has certainly been significantly reduced due to annihilations during the lifetime of the star, but it may still be enhanced compared to the usual NFW case. A detailed calculation of this effect is strongly model-dependent. As we have seen above, the strongest constraints are obtained for direct annihilation into gamma-rays, which is the case we pursue here in more detail.

So far, we assumed that dark stars form in halos between the filtering mass M_F and the mass corresponding to a virial temperature of 10^4 K, M_c. To obtain an upper limit, it is sufficient to assume that in all halos above M_F a dark star remnant will form at some point.

5.5 Cosmic constraints on massive dark matter candidates

Such an assumption clearly overestimates the total contribution at low redshift. When the dark star has formed, a fraction $f_{\mathrm{core}} \sim 10^{-6}$ of the dark matter from the total halo is in the star [Freese et al., 2008a]. For the upper limit, we assume that the total amount of dark matter in star will contribute to the X-ray background (in fact, however, only the dark matter left over in the final remnant can contribute). In this case, we have a proper volume emissivity

$$P_\nu = \delta((1+z)\nu - m_{\mathrm{DM}}) \frac{m_{\mathrm{DM}}}{\mathrm{keV}} \,\mathrm{keV} \alpha_{511} \, f_{\mathrm{r}} f_{\mathrm{a}}$$
$$\times \ n_{\mathrm{DM}} f_{\mathrm{core}} \frac{df_{\mathrm{coll}}(M_F)}{dt}, \tag{5.15}$$

where n_{DM} is the mean proper number density of dark matter particles, m_{DM} the particle mass in keV and df_{coll}/dt can be evaluated from Eq. (5.5). The model-dependent factor f_r determines which fraction of the dark matter in the star will be left in the remnant. We adopt $f_r = 1$ to obtain an upper limit. The factor f_a determines the fraction of the remaining dark matter which actually annihilates, which we set to $f_a = 1$ as well. In Fig. 5.6, we compare the results with EGRET observations [Strong et al., 2004]. We find that the maximum contribution is clearly above the observed background.

Whether this maximum contribution can be reached, is however uncertain and the previous work in the literature only allows one to make rather crude estimates. For instance Iocco et al. [2008] calculate the density profile for a fiducial 100 $M_\odot$ protostar, finding that the density within the star roughly scales with r^{-2} outside a plateau at a radius $r \sim 10^{11}$ cm. At the stellar radius of $\sim 10^{14}$ cm, the dark matter density is still $\sim 10^{12}$ GeV cm^{-3}. The timescale to remove this dark matter enhancement by annihilation is ~ 100 Myr for 100 GeV neutralinos. We need to estimate which fraction of the dark matter inside the star will be left at the end of its life, where the gas density is expelled by a supernova explosion and the dark matter annihilation from this region may contribute to the cosmic gamma-ray background.

Yoon et al. [2008] adopted a timescale of 100 Myr, the typical merger timescale at these redshifts, as the maximum lifetime for dark stars. Depending on the scattering cross section and the environmental density, the actual dark star lifetime may be considerably shorter. Indeed, as we showed in Chapter 5.3.3, it is difficult to reconcile lifetimes of ~ 100 Myr with appropriate reionization scenarios. It is therefore reasonable to assume shorter timescales. In such a case, a reasonable estimate is that $\sim 40\%$ of the dark matter inside the star would be left at the end of its life. This would still be enhanced compared to the standard NFW profile. In this case, the parameter f_r is $\sim 40\%$, and f_a may be of order 1, as the annihilation timescale is comparable to the Hubble time. We note that these numbers are highly uncertain, in particular regarding the exact evolution of dark matter density during the lifetime of the star, the effect of a supernova explosion on the dark matter cusp as well as the consequences of minor mergers.

5. DARK STARS: IMPLICATIONS AND CONSTRAINTS FROM COSMIC REIONIZATION AND EXTRAGALACTIC BACKGROUND RADIATION[1]

There is however also a viable possibility that the dark matter distribution inside the star is significantly steeper than assumed above. In the case of dark matter capture by off-scattering from baryons, the dark matter density inside the star follows a Gaussian shape and is highly concentrated in a small region of $r \sim 2 \times 10^9$ cm [Griest, 1987; Iocco, 2008; Taoso et al., 2008; Yoon et al., 2008]. The implications are not entirely clear. If capture of dark matter stops at the end of the life of the star, the density inside the star will annihilate away quickly, and no significant contribution may come from the remnant. If, on the other hand, dark matter capture goes on until the end of the life of the star, a contribution to the background seems viable. In summary, this may provide a potential contribution to the cosmic background, but its strength is still highly uncertain and should be explored further by future work.

5.5.4 Dependence of dark star models on the neutralino mass

We conclude this section with a discussion on the constraints from cosmic backgrounds for different neutralino masses. As dark star models in the literature mostly consider neutralinos of 100 GeV, there are uncertainties that need to be addressed when considering different neutralino masses. For models involving the capture of dark matter, Iocco [2008] states that the mass of the neutralino does not change the annihilation luminosity. Taoso et al. [2008] find that variations due to different neutralino masses are less than 5%. While these results may hold for high masses, Spergel & Press [1985] showed that for neutralino masses below 4 GeV, they would evaporate from the star, as scattering with baryons can upscatter them as well.

In addition, the AC phase may be modified as well, as the dark matter annihilation rate in this phase is degenerate in the parameter $\langle \sigma v \rangle / m_{\rm DM}$. Iocco et al. [2008] find that the duration of the AC phase may change by almost 50% if the dark matter mass is changed by a factor of 2. The effect of different neutralino masses is therefore uncertain and should be explored in more detail. We will however assume that the general behaviour involving adiabatic contraction in the minihalo is still similar, such that the calculations below are approximately correct also for different neutralino masses.

5.6 Cosmic constraints on light dark matter

Observations of 511 keV emission in the center of our Galaxy [Knödlseder et al., 2003] provide recent motivation to models of light dark matter [Boehm et al., 2004b]. Such observational signatures can be explained assuming dark matter annihilation, while other models still have difficulties reproducing the observations [Boehm et al., 2004a]. The model assumes that dark matter annihilates into electron-positron pairs, which in turn annihilate into 511 keV

5.6 Cosmic constraints on light dark matter

photons. Direct annihilation of dark matter into gamma-rays or neutrinos is assumed to be suppressed to avoid the gamma-ray constraints and to ensure a sufficient positron production rate. It is known that electron-positron annihilation occurs mainly via positronium-formation in our galaxy [Kinzer et al., 2001]. In addition, it was shown [Beacom et al., 2005] that dark matter annihilation to electron-positron pairs must be accompanied by a continuous radiation known as internal bremsstrahlung, arising from electromagnetic radiative corrections to the dark matter annihilation process.

Motivated by these results, it was proposed that internal bremsstrahlung from dark matter annihilation may be responsible for the gamma-ray background at energies of 1-20 MeV [Ahn & Komatsu, 2005b]. Conventional astrophysical sources cannot explain the observed gamma-ray background at these frequencies [Ahn et al., 2005]. A comparison of the observed and predicted background below 511 keV yields constraints on the dark matter particle mass [Ahn & Komatsu, 2005a]. Here we examine whether and how this scenario is affected if dark stars form in the early universe. We use a thermally averaged cross section $\langle \sigma v \rangle \sim 3 \times 10^{-26}$ cm^3 s^{-1} to account for the observed dark matter density [Drees & Nojiri, 1993]. This implies that $\langle \sigma v \rangle$ is velocity-independent (S-wave annihilation). While Boehm et al. [2004a] argue that S-wave annihilation overpredicts the flux from the galactic center, others argue that it is still consistent [Ahn & Komatsu, 2005b,a]. The cross-section adopted here is well-within the conservative constraints of Mack et al. [2008]. The effect of light dark matter annihilation on structure formation in the early universe has been studied in various works, e. g. [Mapelli et al., 2006; Ripamonti et al., 2007a,b]. Constraints from upcoming 21 cm observations have been explored by Furlanetto et al. [2006b] and Valdés et al. [2007], while constraints from background radiation have been considered by Mapelli & Ferrara [2005]. The effects of early dark matter halos on reionization have been addressed recently by Natarajan & Schwarz [2008].

As in the previous section, we point out that significant uncertainties are present when considering dark star models for different dark matter masses, as this question is largely unexplored. In particular, we emphasize that no capturing phase will be present for light dark matter, as shown in the work of Spergel & Press [1985]. Another uncertainty is the question whether to adopt self-annihilating dark matter (i. e. Majorana particles) or particles and antiparticles of dark matter. In the calculations below, we assume that light dark matter is self-annihilating. Otherwise, our results would be changed by a factor of 0.5.

5.6.1 511 keV emission

The expected X-ray background from 511 keV emission is calculated from Eq. (5.8). The volume emissivity of 511 keV photons is given as

$$P_\nu = \delta((1+z)\nu - \nu_{511})\, 511 \text{ keV} \alpha_{511} \langle \sigma v \rangle n_{\text{DM}}^2 C_{511}, \qquad (5.16)$$

5. DARK STARS: IMPLICATIONS AND CONSTRAINTS FROM COSMIC REIONIZATION AND EXTRAGALACTIC BACKGROUND RADIATION[1]

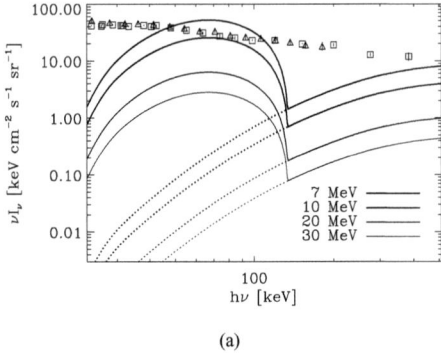

(a)

Figure 5.7: The predicted X-ray background due to 511 keV emission for different dark matter particle masses. Solid lines: Enhanced signal due from adiabatic contraction, dotted lines: Conventional NFW profiles. The observed X-ray background from the HEAO experiments (squares) [Gruber et al., 1999] and Swift/BATSE (triangles) [Ajello et al., 2008] is shown as well. The comparison yields a lower limit of 10 MeV on the dark matter mass for the adiabatically contracted profiles, and 7 MeV for standard NFW halo profiles.

5.6 Cosmic constraints on light dark matter

where $\langle \sigma v \rangle$ denotes the thermally-averaged annihilation cross section, α_{511} is the fraction producing an electron-positron pair per dark matter annihilation process and ν_{511} the frequency corresponding to 511 keV. In our galaxy, this process happens via positronium formation [Kinzer et al., 2001], and we assume that the same is true for other galaxies. In 25% of the cases, positronium forms in a singlet (para) state which decays to two 511 keV photons, whereas 75% form in a triplet (ortho) state which decays into a continuum. We thus adopt $\alpha_{511} = 1/4$ for 511 keV emission. C_{511} refers to the dark matter clumping factor, which is still highly uncertain. The main uncertainty is due to the adopted dark matter profile and the assumptions regarding substructure in a halo [Ahn & Komatsu, 2005a; Ando, 2005; Chuzhoy, 2008; Cumberbatch et al., 2008].

Here we use the clumping factor for a NFW dark matter profile [Navarro et al., 1997] which has been derived by Ahn & Komatsu [2005b,a], as to our knowledge, no calculations of dark star formation are available for other dark matter profiles. For $z < 20$, it is given as a power-law as

$$C_{DM} = C_{DM}(0)(1+z)^{-\beta}, \tag{5.17}$$

where $C_{DM}(0)$ describes the clumping factor at redshift zero and β determines the slope. For a NFW profile [Navarro et al., 1997], $C_{DM}(0) \sim 10^5$ and $\beta \sim 1.8$. The effects of different clumping factors will be explored in future work [Schleicher et al., 2009b]. Ahn & Komatsu [2005b,a] included contributions from all halos with masses above a minimal mass scale M_{min}, which was given as the maximum of the dark matter Jeans mass and the free-streaming mass. This approach assumes instantaneous annihilation of the created electron-positron pairs. As pointed out by Rasera & Teyssier [2006], the assumption of instantaneuos annihilation is only valid if the dark matter halo hosts enough baryons to provide a sufficiently high annihilation probability, postulating this to happen in halos with more than $10^7 - 10^{10} \, M_\odot$, corresponding to their calculation of the filtering mass. We also calculate the filtering mass according to the approach of Gnedin & Hui [1998]; Gnedin [2000], but obtain somewhat lower masses, with $\sim 10^5 \, M_\odot$ halos at the beginning of reionization and $\sim 3 \times 10^7 M_\odot$ at the end [Schleicher et al., 2008b]. This is also in agreement with numerical simulations of Greif et al. [2008] which find efficient gas collapse in halos of $10^5 \, M_\odot$. The discrepancy may also be due to their different reionization model, which assumes reionization to start at redshift 20.

We recall that the clumping factor can be considered as the product of the mean halo overdensity, the fraction of collapsed halos above a critical scale and the mean "halo clumping factor" that describes dark matter clumpiness within a halo. To take into account that electron-positron annihilation occurs only in halos above the filtering mass M_F, we thus rescale the results of Ahn & Komatsu [2005b,a] as

$$C_{511} = \frac{f_{coll}(M_F)}{f_{coll}(M_{min})} C_{DM} f_{enh}, \tag{5.18}$$

5. DARK STARS: IMPLICATIONS AND CONSTRAINTS FROM COSMIC REIONIZATION AND EXTRAGALACTIC BACKGROUND RADIATION[1]

where the factor f_{enh} is given from Eq. (5.13). As above, we assume that the halo mass must be larger than the filtering mass, and lower than the critical mass scale M_c that corresponds to virial temperatures of 10^4 K. For comparison, we will also calculate 511 keV emission with $f_{enh} = 1$. We note that the resulting background will be somewhat lower than the result of Ahn & Komatsu [2005a], as we adopted $\alpha_{511} = 1/4$ and include only halos above the filtering mass scale in the clumping factor. In Fig. 5.7, we compare the results with the observed X-ray background from the HEAO-experiments [1] [Gruber et al., 1999] and SWIFT [2]/BATSE [3] observations [Ajello et al., 2008]. In the standard NFW case, we find a lower limit for the dark matter particle mass of 7 MeV. For the case with adiabatically contracted profiles due to dark star formation, we find a slightly higher lower limit of 10 MeV. This is because the enhancement is effective only for frequencies $h\nu < 100$ keV, where the observed background is significantly larger than at 511 keV, where Ahn & Komatsu [2005a] obtained their upper limit.

5.6.2 Internal Bremsstrahlung

The internal bremsstrahlung is calculated according to the approach of Ahn & Komatsu [2005b]. The background intensity is given by Eq. (5.8), with a proper volume emissivity

$$P_\nu = \frac{1}{2} h\nu \langle \sigma v \rangle C_{brems} n_{DM}^2 \left[\frac{4\alpha}{\pi} \frac{g(\nu)}{\nu} \right], \quad (5.19)$$

where $\alpha \sim 1/137$ is the finestructure constant and $g(\nu)$ is a dimensionless spectral function, defined as

$$g(\nu) = \frac{1}{4} \left(\ln \frac{\tilde{s}}{m_e^2} - 1 \right) \left[1 + \left(\frac{\tilde{s}}{4m_{DM}^2} \right)^2 \right], \quad (5.20)$$

with $\tilde{s} = 4m_{DM}(m_{DM} - h\nu)$. As Ahn & Komatsu [2005b] pointed out in a 'Note added in proof', bremsstrahlung is emitted in all dark matter halos, regardless of the baryonic content. There is thus no need to consider any shift in the minimal mass scale, the only thing to take into account is the enhancement of annihilation due to the AC profiles. The clumping factor C_{brems} is thus given as

$$C_{brems} = C_{DM} f_{enh}, \quad (5.21)$$

where f_{enh} is given by Eq. (5.8). In Fig. 5.8, we compare the results with the observed gamma-ray background from the HEAO-experiments [Gruber et al., 1999] and SWIFT/BATSE observations [Ajello et al., 2008], as well as SMM [4] [Watanabe et al., 1999] and Comptel

---

[1] http://heasarc.gsfc.nasa.gov/docs/heao1/heao1.html
[2] http://heasarc.nasa.gov/docs/swift/swiftsc.html
[3] http://www.batse.msfc.nasa.gov/batse/
[4] http://heasarc.gsfc.nasa.gov/docs/heasarc/missions/solarmax.html

5.6 Cosmic constraints on light dark matter

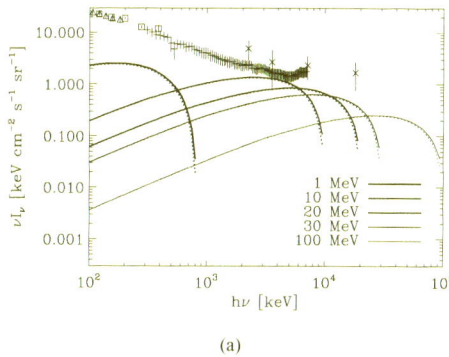

(a)

Figure 5.8: The predicted gamma-ray background due to bremsstrahlung emission for different dark matter particle masses. Solid lines: Enhanced signal due from adiabatic contraction, dotted lines: Conventional NFW profiles. The lines overlap almost identically, as the main contribution comes from redshift zero, where the clumping factor is large and dark stars are assumed not to form. The observed gamma-ray background from the HEAO experiments (squares) [Gruber et al., 1999], Swift/BATSE (triangles) [Ajello et al., 2008], COMPTEL (crosses) [Kappadath et al., 1996] and SMM (plusses) [Watanabe et al., 1999] is shown as well.

5. DARK STARS: IMPLICATIONS AND CONSTRAINTS FROM COSMIC REIONIZATION AND EXTRAGALACTIC BACKGROUND RADIATION[1]

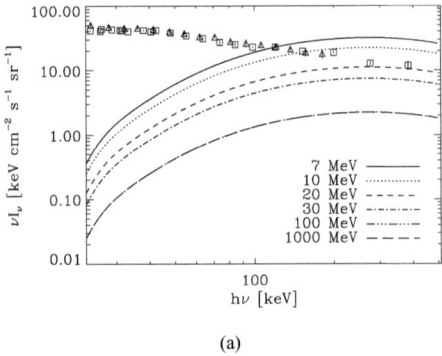

(a)

Figure 5.9: The upper limit of X-ray radiation due to dark star remnants. The observed X-ray background from the HEAO experiments (squares) [Gruber et al., 1999] and Swift/BATSE (triangles) [Ajello et al., 2008] is shown as well. Only for very low dark matter particle masses, the upper limit is somewhat higher than the observed background. However, the actual contribution may be lower by some orders of magnitude (see discussion in the text).

---

[1] data [Kappadath et al., 1996]. We find that the signal is almost unchanged in the model taking into account dark star formation. The reason is that dark stars form mainly at high redshifts, in the range where $M_F < M_c$, while the dominant contribution to the background comes from redshift zero. Our results agree with Ahn & Komatsu [2005b].

5.6.3 Emission from dark star remnants

As in Chapter 5.5.3, we consider a scenario where dark stars explode at the end of their lifetime and dark matter annihilation products in their remnants contribute to the cosmic background. For simplicity, we consider 511 keV emission only, which is also mostly sensitive to modifications at high redshift. Again, we assume that dark stars form in halos between the filtering mass M_F and the mass corresponding to a virial temperature of 10^4 K, M_c. In this case, the volume emissivity is given as

$$\begin{aligned} P_\nu &= \delta\left((1+z)\nu - \nu_{511}\right) 511 \text{ keV} \alpha_{511}\, f_r f_a \\ &\quad \times n_{\text{DM}} f_{\text{core}} \frac{df_{\text{coll}}(M_F)}{dt}, \end{aligned} \qquad (5.22)$$

---

[1] http://wwwgro.unh.edu/comptel/

where n_{DM} is the mean proper number density of dark matter particles and df_{coll}/dt can be evaluated from Eq. (5.5). The model-dependent factor f_r determines which fraction of the dark matter in the star will be left in the remnant, we adopt $f_r = 1$ to obtain an upper limit. The factor f_a determines the fraction of the remaining dark matter which actually annihilates, which we set to $f_a = 1$ as well. As in Chapter 5.6.1, $\alpha_{511} = 1/4$ is the fraction of electron-positron annihilations per one dark matter annihilation process, corresponding to annihilation via positronium formation. In Fig. 5.9, we compare the results with the observed X-ray background from the HEAO-experiments [Gruber et al., 1999] and SWIFT/BATSE observations [Ajello et al., 2008].

For dark matter particle masses below 30 MeV, the upper limit found here is higher than the observed background. Again, as discussed in Chapter 5.5.4, there are significant uncertainties regarding the question whether this high contribution can be reached, both due to uncertainties in the dark star models, which have not been explored for light dark matter, as well as the impact of a supernova explosion on the dark matter cusp. These possibilities should be addressed further in future work.

5.7 Summary and discussion

In this work, we have examined whether the suggestion of dark star formation in the early universe is consistent with currently available observations. We use these observations to obtain constraints on dark star models and dark matter properties. From considering cosmic reionization, we obtain the following results:

- Dark stars with masses of the order 800 $M_\odot$ as suggested by Freese et al. [2008a] can only be reconciled with observations if somewhat artificial double-reionization scenarios are constructed. They consist of a phase of dark star formation followed by a phase of weak Pop. II star formation and a final star burst to reionize the universe until redshift 6.

- The same is true for dark stars in which the number of UV photons is significantly increased due to dark matter capture, as suggested by Iocco et al. [2008].

- It appears more reasonable to require that dark stars, if they were common, should have similar properties as conventional Pop. III stars. For MS-dominated models, this requires that typical dark star masses are of order 100 $M_\odot$ or below. For CD models it requires a dark matter density above $10^{11} - 10^{12}$ GeV cm^{-3} if a spin-dependent elastic scattering cross section of $t \times 10^{-39}$ cm^2 is assumed [Yoon et al., 2008].

5. DARK STARS: IMPLICATIONS AND CONSTRAINTS FROM COSMIC REIONIZATION AND EXTRAGALACTIC BACKGROUND RADIATION[1]

- Alternatively, it may imply that the elastic scattering cross section is smaller than the current upper limits, that the dark matter cusp is destroyed by mergers or friction with the gas or that the star is displaced from the center of the cusp.

- A further interpretation is that dark stars are very rare. This would require some mechanism to prevent dark star formation in most minihalos.

- However, if the double-reionization models are actually true, it would indicate that dark stars form only at redshifts beyond 14, which makes direct observations difficult.

- We also note that 21 cm observations may either confirm or rule out double-reionization models.

We have also examined whether the formation of dark stars and the corresponding enhancement of dark matter density in dark matter halos due to adiabatic contraction may increase the observed X-ray, gamma-ray and neutrino background. Here we found the following results:

- For massive dark matter particles, direct annihilation into gamma-rays provides significant constraints for masses less than 30 GeV.

- For massive dark matter particles, the contribution from direct annihilation into neutrinos is well below the observed background.

- In light dark matter scenarios, the 511 keV emission is significantly enhanced below frequencies of 100 keV in the observers restframe. For a certain range of parameters, this emission may even form a significant contribution of the total X-ray background. In this case, we derive a lower limit of 10 MeV for the dark matter particle mass (while we find 7 MeV for standard NFW profiles).

- In light dark matter scenarios, the background radiation due to internal bremsstrahlung is not affected significantly from adiabatic contraction at early times, as the main contribution comes from low redshift.

- Both for light and massive dark matter particles, the annihilation products in the remnants of dark stars may provide significant contributions that may be used to constrain such models in more detail. However, whether this contribution can be reached is highly model-dependent and relevant questions regarding the death of dark stars has not been explored in the literature.

5.7 Summary and discussion

Future observations may provide further constraints on this exciting suggestion. Small-scale 21 cm observations may directly probe the HII regions of the first stars and provide a further test of the luminous sources at high redshift, and extremely bright stars might even be observed with the James-Webb telescope, if they form sufficiently late. With this work, we would like to initiate a discussion on observational tests and constraints on dark stars, which may tighten theoretical dark star models and provide a new link between astronomy and particle physics.

5. DARK STARS: IMPLICATIONS AND CONSTRAINTS FROM COSMIC REIONIZATION AND EXTRAGALACTIC BACKGROUND RADIATION[1]

6

Cosmic constraints rule out s-wave annihilation of light dark matter[1]

In this chapter[1], it is shown that light dark matter models assuming a constant thermally-averaged annihilation cross section (s-wave annihilation) can be ruled out by a combination of cosmic and galactic constraints.

6.1 Introduction

Recent observations of the Galactic center in different frequencies have provided increasing motivation for theoretical models that consider dark matter annihilation and/or decay. One finds an excess of GeV photons [de Boer et al., 2005], of microwave photons [Hooper et al., 2007], of positrons [Cirelli et al., 2008] and of MeV photons [Jean et al., 2006; Weidenspointner et al., 2006], which correlates with the Galactic bulge instead of the disk and cannot be attributed to single sources. It is therefore controversial whether it can be explained by conventional astrophysical sources alone or if dark matter annihilation models are required. It is not clear whether these different phenomena are related. Hence, their interpretation is still under discussion [de Boer, 2008]. For this reason, it is particularly interesting to also consider constraints which are independent of the Galactic center observations.

Observations in the MeV energy range favor models based on light dark matter. These models suggest that light dark matter particles with masses of 1 – 100 MeV annihilate into electron-positron pairs [Boehm et al., 2004a]. In such a scenario, the dark matter mass needs to be larger than 511 keV to be able to produce electron-positron pairs through dark matter annihilation. It should be smaller than 100 MeV, as otherwise pion final states that produce

---

[1]The work presented here is based on an article that is reprinted with permission from Schleicher, Glover, Banerjee & Klessen, PRD, 79, 023515, 2009. Copyright (2009) by the American Physical Society.

6. COSMIC CONSTRAINTS RULE OUT S-WAVE ANNIHILATION OF LIGHT DARK MATTER[1]

too many gamma rays would be possible [Boehm et al., 2004a]. Electromagnetic radiative corrections to the annihilation process require the emission of internal bremsstrahlung [Beacom et al., 2005]. It has also been proposed that such internal bremsstrahlung emission might explain the observed gamma-ray background in the 10 – 20 MeV range [Ahn et al., 2005; Ahn & Komatsu, 2005b] for dark matter masses of ∼ 20 MeV, though these models appear less favorable in light of stronger upper limits on the dark matter particle mass [Beacom & Yüksel, 2006; Sizun et al., 2006]. Supernovae data require dark matter particle masses larger than 10 MeV, although this limit depends on assumptions made regarding the scattering cross section between dark matter particles and neutrinos [Fayet et al., 2006].

Calculations regarding the annihilation of light dark matter in the Milky Way show that predicted and observed fluxes are only in agreement for p-wave annihilation models [Boehm et al., 2004a]. In addition, dark matter models can be constrained by the cosmic backgrounds [Ahn & Komatsu, 2005a; Ando, 2005; Rasera & Teyssier, 2006; Yüksel et al., 2007; Mack et al., 2008; Schleicher et al., 2008a]. Such constraints provide a highly complementary approach based on the observed cosmic gamma-ray background. In this letter, we show that such constraints provide a strong independent confirmation that s-wave annihilation of light dark matter is ruled out.

6.2 Assumptions

To explain the observed abundance of dark matter in the universe requires a thermally-averaged annihilation cross-section of [Boehm et al., 2004b; Drees & Nojiri, 1993]

$$\langle \sigma v \rangle \sim 3 \times 10^{-26} \text{ cm}^3 \text{ s}^{-1}. \tag{6.1}$$

In the mass range considered here, this may vary only by 10% [Ahn & Komatsu, 2005b; Boehm et al., 2004b]. Such a cross section is in agreement with conservative constraints derived from gamma-ray observations of the Milky Way, Andromeda (M31) and the cosmic background [Mack et al., 2008]. We adopt Eq. (6.1) in this study.

The overall intensity of annihilation radiation, however, depends sensitively on the dark matter clumping factor. This quantity, defined as $C(z) = \langle \rho_{DM}^2(z) \rangle / \langle \rho_{DM}(z) \rangle^2$, has been subject to several studies [Ahn & Komatsu, 2005a; Ando, 2005; Chuzhoy, 2008; Cumberbatch et al., 2008] and turns out to be highly uncertain. For example, a very detailed study by Cumberbatch et al. [2008] computed clumping factors for three commonly-adopted dark matter halo profiles: Navarro-Frenk-White (NFW) [Navarro et al., 1997], Moore [Moore et al., 1999] and Burkert profiles [Burkert, 1995]. Depending on model assumptions, the Burkert profiles, which have a flat central core, yield clumping factors of the order $10^5 - 3 \times 10^6$, the Moore profiles, which have a steep central cusp, yield factors in the range $10^7 - 10^{11}$, and intermediate NFW profiles yield factors of $10^6 - 3 \times 10^9$ at redshift zero. The large variation in values for a given profile is due to the dependence of the clumping factor on further properties like the amount of substructure and typical halo concentration parameter. There has been a long controversy between different theoretical and observational studies, with observations typically favoring shallow profiles and theoretical investigations favoring power-law behavior with central slopes of the order -1 [Flores & Primack, 1994; Burkert, 1995; Fukushige & Makino, 1997, 2003; Moore et al., 1999; Salucci & Burkert, 2000; Ghigna et al., 2000; Subramanian, 2000; Taylor & Navarro, 2001; Jing & Suto, 2000, 2002; Klypin et al., 2001; de Blok & Bosma, 2002; Ricotti, 2003; Gentile et al., 2004; Tasitsiomi et al., 2004; Hoekstra et al., 2004; Broadhurst et al., 2005]. More recently, there appears some convergence towards the Einasto profile [Einasto, 1965] with a slightly shallower slope of ~ -0.8 [Graham et al., 2006; Springel et al., 2008a; Navarro et al., 2008]. High-resolution simulations further indicate that substructure will not provide a strong contribution to the amount of dark matter annihilation [Springel et al., 2008b], though more realistic simulations including baryonic physics will be needed to finally resolve this issue. As the slope of the Einasto profile is very close to the NFW, clumping factors should be in the same range, and certainly above the flat Burkert profile. This favors clumping factors in the range $10^6 - 10^7$, and as a firm lower limit, we adopt 10^5 at redshift zero, which is the lowest value found for calculations with the Burkert profile, and a factor of 30 below the lowest value for the NFW profile [Cumberbatch et al., 2008]. We note that some works in the literature adopt an even smaller minimal clumping factor of 2×10^4 [Ando, 2005; Yüksel et al., 2007; Mack et al., 2008], corresponding to the Kravtsov profile [Kravtsov et al., 1998]. Such a choice would leave our conclusions unchanged.

6.3 Formalism

The gamma-ray background intensity is given as [Ahn & Komatsu, 2005b]

$$I_\nu = \frac{c}{4\pi} \int \frac{dz P_\nu([1+z]\nu, z)}{H(z)(1+z)^4}, \tag{6.2}$$

where c is the speed of light, z the redshift, P_ν the proper volume emissivity and $H(z)$ the Hubble function. The proper volume emissivity is given as

$$P_\nu = \frac{1}{2} h\nu \langle \sigma v \rangle C_{\text{clump}} n_{\text{DM}}^2 \left[\frac{4\alpha}{\pi} \frac{g(\nu)}{\nu} \right], \tag{6.3}$$

where h is Planck's constant, C_{clump} is the dark matter clumping factor, n_{DM} the number density of dark matter particles, $\alpha \sim 1/137$ the fine-structure constant and $g(\nu)$ is a dimensionless

6. COSMIC CONSTRAINTS RULE OUT S-WAVE ANNIHILATION OF LIGHT DARK MATTER[1]

spectral function, defined as

$$g(\nu) = \frac{1}{4}\left(\ln\frac{\tilde{s}}{m_e^2} - 1\right)\left[1 + \left(\frac{\tilde{s}}{4m_{DM}^2}\right)^2\right], \quad (6.4)$$

with $\tilde{s} = 4m_{DM}(m_{DM} - h\nu)$. The number density of dark matter particles is calculated from the mass density $\rho_{DM} = m_{DM} n_{DM}$, which is highly constrained from WMAP observations [Komatsu et al., 2008]. As the thermally-averaged cross section is fixed by Eq. (6.1), the predicted gamma-ray background depends only on the assumed dark matter particle mass m_{DM}. Between redshifts 10 and zero, which yield the dominant contribution to the predicted gamma-ray background, the clumping factors are well described by a power-law

$$C_{\text{clump}} = C_0 (1 + z)^{-\beta} \quad (6.5)$$

with $\beta \sim 3$ and C_0 being the clumping factor at redshift zero. As discussed above, the value of C_0 should be in the range $10^6 - 10^7$, and certainly above 10^5. If the dark matter density in virialized halos can be assumed constant, the clumping factor rises as $(1 + z)^{-3}$ towards lower redshift [Chuzhoy, 2008]. The evolution is however less steep if substructure in the halo is neglected [Ahn & Komatsu, 2005b,a]. Such a case would strengthen the constraint given below, as even more radiation will be emitted at high redshift.

6.4 Results

For different dark matter masses, we calculate which clumping factor would be required to exceed the observed gamma-ray background due to emission of internal bremsstrahlung. For illustrative purposes, we give some examples in Fig. 6.1. In Fig. 6.2, we show the constraint for different dark matter masses larger than 511 keV. Further limits can be calculated from 511 keV line emission. As a conservative choice, we assume that the electron-positron-pairs annihilate via positronium formation as observed in the Milky Way [Kinzer et al., 2001], such that only 25% of all annihilations lead to the emission of 511 keV photons. For masses of 7 MeV, one finds a maximum permitted clumping factor of $C_0 = 10^5$ [Schleicher et al., 2008a]. As this emission always peaks at the same frequency, one can obtain the constraint for other particle masses by requiring that $C_0/m_{DM}^2 = \text{const}$. For dark matter masses larger than 11 MeV, the constraint from internal bremsstrahlung is most stringent. The limit from 511 keV emission is only important for dark matter masses below 11 MeV. For masses below the upper limit from Galactic center observations, the constraint from 511 keV emission is strongest.

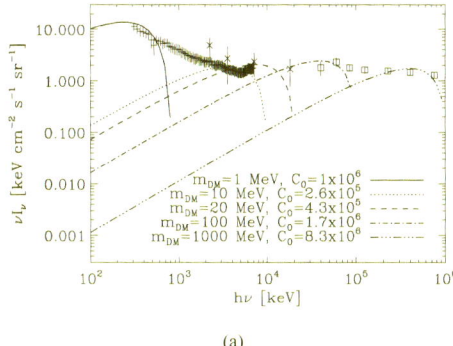

Figure 6.1: The predicted gamma-ray background due to internal bremsstrahlung emission for different dark matter particle masses. In every case, we adopted a clumping factor that yields the maximum allowed background. We compare with the observed gamma-ray background from COMPTEL (crosses) [Kappadath et al., 1996], SMM (plusses) [Watanabe et al., 1999] and EGRET (squares) [Strong et al., 2004].

6.5 Implications and discussion

As shown by Beacom & Yüksel [2006], light dark matter scenarios that explain observations of the Galactic center require a mass less than 3 MeV. This results from a comparison between 511 keV emission and higher-energy gamma-rays and does not require assumptions regarding the dark matter profile. For low masses, clumping factors C_0 in the range $10^6 - 10^7$, which we would expect for the currently-favored Einasto profiles, are excluded due to the gamma-ray background constraints. Even if we adopted clumping factors corresponding to a flat Burket profile with little substructure, corresponding to values $C_0 \geq 10^5$ [Cumberbatch et al., 2008], this would violate the constraints from the gamma-ray background. This shows that the light dark matter model can no longer be maintained with a constant thermally-averaged annihilation cross section if constraints from the cosmic background are combined with the upper mass limits from the spectral constraints, thus providing an independent confirmation that s-wave annihilation of light dark matter can be ruled out.

These conclusions hold even if we adopted the very conservative upper mass limit of Sizun et al. [2006], which is 7.5 MeV if the ISM in the Galactic center would be strongly ionized. This appears unlikely and provides a very strong upper limit. In addition, more recent INTEGRAL observations favor an even smaller emission region, which increases the

6. COSMIC CONSTRAINTS RULE OUT S-WAVE ANNIHILATION OF LIGHT DARK MATTER[1]

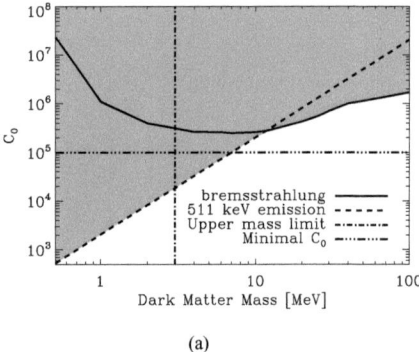

(a)

Figure 6.2: The constraint on the clumping factor C_0 at redshift zero due to internal bremsstrahlung emission. For comparison, we show the weaker constraint due to 511 keV emission. The forbidden region is shaded. We also show the upper mass limit from Galactic center observations calculated by Beacom & Yüksel [2006] as well as the minimal clumping factor $C_0 = 10^5$. The combination of these constraints shows that light dark matter models with s-wave annihilation are ruled out.

6.5 Implications and discussion

inflight annihilation intensity by a factor of 1.7 and pushes the allowed maximum positron injection energy even further down (John Beacom, private communication).

Still feasible are models based on p-wave annihilation, in which the annihilation cross section becomes much smaller at late times when the velocities of dark matter particles are no longer relativistic [Bœhm & Fayet, 2004]. It is also interesting to note that the gamma-ray background at 10 – 20 MeV might be explained by a power-law component of non-thermal electrons in active galactic nuclei (AGN) [Inoue et al., 2008]. It is however unclear whether a sufficient number of non-thermal electrons is actually available. We expect that the FERMI satellite [1] will shed more light on such questions, as the anisotropic distribution of the gamma-ray background may allow one to distinguish between astrophysical sources and dark matter annihilation [Ando et al., 2007].

We finally note that we have checked if similar constraints apply for more massive dark matter candidates, based on the recent analysis of Mack et al. [2008] and Yüksel et al. [2007]. For such models, however, clumping factors in the range $10^7 - 10^9$ are still feasible and well within the allowed parameter space. The upper limits derived for light dark matter do not apply here, as the annihilation products may be different.

We finally note that if the clumpiness of dark matter was high already at early times, annihilation of light dark matter may provide a significant contribution to the observed reionization optical depth [Schleicher et al., 2008b; Chuzhoy, 2008; Natarajan & Schwarz, 2008]. The detailed effects of such annihilations on the IGM have been explored by Ripamonti et al. [2007a], and consequences for 21 cm observations have been explored by Furlanetto et al. [2006b]; Chuzhoy [2008].

---

[1] http://www.nasa.gov/mission_pages/GLAST/science/index.html

6. COSMIC CONSTRAINTS RULE OUT S-WAVE ANNIHILATION OF LIGHT DARK MATTER[1]

7

Probing high-redshift quasars with ALMA[1]

In this chapter[1], it is shown how ALMA opens a new observational window to probe the chemical and physical properties of high-redshift quasars via molecular and fine structure lines. Particular focus is on the signatures of X-ray emission from the central black hole the discrimination from the starburst. This is an extension of a previous study by Spaans & Meijerink [2008], who calculated observables for ALMA in low-metallicity gas in active galaxies at $z \sim 15$. Here, we focus on the solar metallicity case, and predict the expected fluxes from the central X-ray dominated region in CO, [CII] and [OI].

7.1 Introduction

The recent detection of kpc-scale star-forming structures at $z = 6.42$ through the detection of [CII] emission [Walter et al., 2009a] and emission in various rotational CO lines [Riechers et al., 2009] confirms the importance of mm/sub-mm observations to infer gas distribution and dynamics in quasar host galaxies. Emission in CO and the continuum in high-redshift quasars have also been reported by Omont et al. [1996], Carilli et al. [2002], Walter et al. [2004], Klamer et al. [2005], Weiß et al. [2005], Maiolino et al. [2007], Walter et al. [2007], Weiß et al. [2007], Riechers et al. [2008b] and Riechers et al. [2008a].

With the upcoming mm/sub-mm telescope ALMA[2], it will be possible to probe such structures in even more detail, due to its significantly improved sensitivity, angular and spectral resolution. From a theoretical point of view, it is therefore interesting to speculate what

---

[1]This chapter is based on an article by Schleicher, Spaans & Klessen, A&A, 513, 7, 2010, reproduced with permission © ESO.
[2]http://www.eso.org/sci/facilities/alma/

7. PROBING HIGH-REDSHIFT QUASARS WITH ALMA[1]

ALMA might observe in the center of such host galaxies. Within the next ten years, we further expect the advent of SPICA[1] and FIRI[2], which will probe the universe in the mid- or far-IR regime, respectively. These telescopes will be able to apply in the local universe what we suggest as high-redshift diagnostics for ALMA.

Before speculating what ALMA may see in the centers of high-redshift galaxies, we turn our attention to the properties of molecular clouds in the central molecular zone (CMZ) of the Milky Way. Studies employing H_3^+ and CO lines indicate the presence of high-temperature ($T \sim 250$ K) and low-density ($n \sim 100$ cm^{-3}) gas [Oka et al., 2005]. NH$_3$ observations by Nagayama et al. [2007] confirm the presence of warm molecular clouds, with temperatures mostly between 20 – 80 K. The presence of a 120 pc star-forming ring was infered by CO observations, indicating typical densities of $10^{3.5-4}$ cm^{-3} and kinetic temperatures of 20 – 35 K [Nagai et al., 2007]. These temperatures were derived under the conservative assumption of a beam filling factor 1, while smaller filling factors would give rise to higher temperatures.

In the centers of active galaxies, the supermassive black hole will emit radiation in a broad range of frequencies. Particularly interesting is the emission of X-rays, as those photons can penetrate molecular clouds even at high column densities. The resulting cloud temperatures range from a hundred K up to 1000 K [Lepp & Dalgarno, 1996; Maloney et al., 1996; Meijerink & Spaans, 2005; Meijerink et al., 2007]. Such X-ray dominated regions have been reported for instance in NGC 1068 by Galliano et al. [2003]. Due to the high ambient pressure, higher-density clouds may form due to the thermal instability [Wada & Norman, 2007].

At high redshift, resolution of present-day telescopes is generally not sufficient to resolve the central regions of quasar host galaxies. Under exceptional circumstances, this is however possible. Indeed, highly excited high-J CO and HCN line emission was found in APM 08279+5255 by Weiß et al. [2007]. As this galaxy is gravitationally lensed, it was possible to measure fluxes from the central X-ray dominated region that are usually beam-diluted. While the CO line fluxes usually rise up to the CO (5-4) transition and then decrease, they continue to rise in this system up to the CO (10-9) transition, providing clear evidence for the presence of warm gas in the central region. Similar results have also been obtained for the Cloverleaf quasar [Bradford et al., 2009].

Motivated by these results, we study in more detail the possibility to probe high-redshift quasar host galaxies with ALMA, and in particular their central regions. In § 7.2, we review the main heating mechanisms that may be present in high-redshift quasars, and discuss their influence on the chemistry. In § 7.3, we give a brief summary on the main PDR observables, which are already used to probe galaxies at high redshift. As a specific example, we calculate the expected fluxes for the Seyfert 2 galaxy NGC 1068 if it were located at $z = 8$. Our expectations for the central X-ray dominated regions (XDRs) are formulated in § 7.4 based on

---
[1] http://www.ir.isas.jaxa.jp/SPICA/index.html
[2] http://sci.esa.int/science-e/www/object/index.cfm?fobjectid=40090

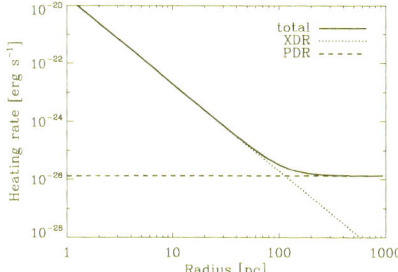

Figure 7.1: The heating rates per hydrogen atom due to X-ray absorption (XDR contribution), absorption of soft UV-photons (PDR contribution) and total heating rate as a function of radius. The calculation assumes a 10^7 $M_\odot$ black hole, with 3% of its Eddington luminosity being emitted in a hard spectral component between 1 and 100 keV with a spectral slope of -1. The strength of the soft UV radiation field is taken as $G_0 = 10$ in Habing units and the effective density $n_{\rm eff} = 10^5$ cm^{-3}. For such a configuration, the XDR contribution clearly dominates within the central 100 pc.

detailed chemical models including ~ 50 species and several thousand reactions. Evidence for X-ray dominated regions at different redshifts is reviewed and discussed in § 7.5. On this basis, we assess the prospects for finding new sources in an ALMA deep field in § 7.6. We conclude in § 7.7. In summary, this paper provides a basic set of predictions concerning observations of high-redshift quasars with ALMA. In a companion paper, we plan to provide diagnostics based on the observed line fluxes that will allow one to infer physical properties such as the star formation rate or the X-ray luminosity based on the observed line emission.

7.2 Chemistry in high-redshift quasars

Emission from molecular clouds in active galaxies can be excited by a variety of different mechanisms. Mechanical feedback may be important locally, in particular in the presence of shocks or outflows [Loenen et al., 2008; Papadopoulos et al., 2008]. In addition, there is radiation in a broad range of frequencies, both from the starburst and the supermassive black hole. UV emission generally gives rise to compact HII regions at temperatures of $\sim 10^4$ K in which molecules are completely dissociated and emission is mostly by Lyman α and various fine-structure lines. Such photons are however absorbed by relatively small column densities, and stellar HII regions never become larger than a few pc, considerably smaller than the scales of interest here.

7. PROBING HIGH-REDSHIFT QUASARS WITH ALMA[1]

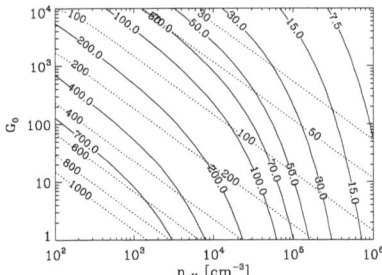

Figure 7.2: The expected size of the X-ray dominated region in pc, for a black hole with $10^7\,M_\odot$ with a spectral slope of -1, as a function of the soft UV radiation field G_0 in Habing units and the effective number density n_{eff}. The straight lines assume the power-law between 1 and 5 keV, while the dotted lines assume it between 1 and 100 keV.

Soft UV-photons have smaller cross sections and may penetrate larger columns. In the presence of a starburst, they are therefore the dominant driver of the molecular cloud chemistry [Hollenbach & Tielens, 1999]. While their heating efficiency is low (0.1 − 0.3%), they are very efficient in dissociating molecules. In such photon-dominated regions (PDRs), one generally expects somewhat enhanced cloud temperatures, with most of the emission in fine-structure lines, as CO is efficiently dissociated. A detailed review of these processes is given by Meijerink & Spaans [2005].

X-ray photons have even smaller cross sections than the soft-UV photons, and can thus penetrate larger columns. Specifically, a 1 keV photon penetrates a typical column of 2×10^{22} cm^{-2}, a 10 keV photon penetrates 4×10^{25} cm^{-2} and a 100 keV photon 9×10^{30} cm^{-2}. For this reason, X-rays can keep molecular clouds at high temperatures even at high column densities. They have high heating efficiencies of the order of 30%, and are inefficient in the dissociation of molecules. A fraction of them may however be reprocessed by the gas and converted in soft-UV photons, which may lead to some molecular dissociation. Detailed reviews are given by Maloney et al. [1996]; Lepp & Dalgarno [1996]; Meijerink & Spaans [2005]. Therefore, X-ray absorption drives a completely different type of chemistry, and may potentially result in temperatures up to 1000 K. The fraction of molecules, in particular CO, can be very high.

To estimate the potential extent of such a central X-ray dominated region, we employ a toy model that takes into account the heat input from the starburst and from the X-ray emission of the supermassive black hole. We assume here an axisymmetric situation with a central supermassive black hole and an extended molecular disk in the host galaxy. The radiation from the SMBH will consist of a soft and a hard component. The soft component is easily absorbed at the edge of the molecular clouds and can be neglected for column densities of 10^{22} cm^{-2} and above. The X-ray photons, on the other hand, penetrate deeply into the molecular disk and excite emission there. For the X-ray photons, we adopt a power-law spectrum for frequencies larger than 1 keV and use the cross sections given by Verner & Yakovlev [1995]. For soft photons from the starburst, we adopt a typical frequency of 10 eV and a cross section of 2.78×10^{-22} cm^{-2} [Meijerink & Spaans, 2005]. Typical heating efficiencies are 30% in XDRs and 0.3% in PDRs. We assume that the radiation field from the starburst is roughly constant within the central region. The X-ray radiation field, on the other hand, will be geometrically diluted and partially shielded by the gas. To calculate the attenuation of X-rays, we introduce an effective density for the central region, which is given as

$$n_{\text{eff}} = \alpha 10^5 \text{ cm}^{-3} + (1-\alpha)10^2 \text{ cm}^{-3}, \quad (7.1)$$

where α is the volume-filling factor of dense clouds, 10^5 cm^{-3} is a typical cloud density and 10^2 cm^{-3} a typical density of the atomic medium. We adopt now a specific reference case for which we evaluate the heating rates and the expected size of the XDR. The X-ray emission depends essentially on the product of black hole mass M_{BH}, the Eddington ratio λ_E, and the fraction f_X of the total luminosity going into the hard spectrum. We assume a modest black hole mass $M_{BH} = 10^7\ M_\odot$, an Eddington ratio $\lambda_E = 30\%$, which lies in the typical range of measured Eddington ratios for high-redshift AGN [Shankar et al., 2004; Kollmeier et al., 2006; Shankar et al., 2009], and a fraction of $f_X = 10\%$ of the total luminosity emitted in X-rays. For the X-ray spectrum, we adopt a frequency range between 1 and 100 keV with a spectral slope of -1.

We further assume that the starburst produces a soft UV-radiation field of $G_0 = 10$ (Habing units) and an effective density of $n_{\text{eff}} = 10^5$ cm^{-3}. Such an effective density corresponds to a volume-filling factor of order 1 and is therefore an upper limit. In real AGN, the effective density may be smaller in the central region, implying less attenuation of the X-rays and therefore a larger X-ray dominated region. On the other hand, the parameter G_0 may vary as well and depend on the strength of the ongoing starburst. For this scenario, the expected heating rates per hydrogen atom are given in Fig. 7.1.

To explore the parameter dependence in more detail, we now consider the effective density and the strength of the starburst radiation field as free parameters and check how they influence the size of the XDR, keeping black hole mass, Eddington ratio and the luminosity fraction in the hard component as specified above. We consider two cases, one with the hard component between 1 and 5 keV (case A), and one with the hard component between 1 and 100 keV (case B). The results for the XDR size are given in Fig. 7.2. In case A shielding effects can be clearly recognized and the size of the XDR depends more on the effective density than on the strength of the starburst. As typical molecular cloud densities in these environments are $\sim 10^5$ cm^{-3} and the filling factor is of order 1%, we can expect an average

7. PROBING HIGH-REDSHIFT QUASARS WITH ALMA[1]

density of 10^3 cm^{-3}. At its largest baseline, ALMA can even resolve spatial scales of $\sim$ 30 pc at $z = 5$, and should thus resolve the corresponding XDRs.

Simulations by Wada et al. [2009] for the clumpy medium at scales of $\sim$ 30 pc indicate a volume-filling factor $\alpha \sim 0.03$. Models by Galliano et al. [2003] for NGC 1068, on scales of a few hundred parsec, indicate a volume-filling factor of 0.01, which still leads to a surface-filling factor of order 1. For the central 200 pc of our galaxy, values of $\alpha \sim 0.1 - 0.01$ have been suggested [McCall et al., 1999; Oka et al., 2005]. For a large volume-filling factor, we expect a somewhat smaller XDR due to the attenuation of X-rays. At the same time, however, this region will consist of a large number of clouds that are highly excited. For smaller clumping factors, the number densitiy of clouds will be reduced, but the size of the XDR increased.

Of course, the approach used here is only approximate, as the attenuation may depend on the direction and be stronger in some directions and weaker in others. However, these order-of-magnitude estimates should still be applicable for a broad range of conditions and also hold in cases of spherical rather than flattened structures. An implicit assumption of the model is that we average over sufficiently large scales where the effective density provides a good approximation with respect to X-ray attenuation. For more clumpy structures, the XDR would be more inhomogeneous and reach out further along the low-density regions. In case the large-scale clumpiness is considerable, so that molecular clouds do no longer fill the projected surface area in the beam, our model predictions need to be corrected with the corresponding area-filling factor.

7.3 Observables in the PDR

In photon-dominated regions, soft UV-photons from the starburst provide some heat input for molecular clouds, in particular at low column densities, and are efficient in dissociating molecules like H_2 or CO. The main coolants in this regime are therefore fine-structure lines of [CII] and [OI]. Depending on the strength of the radiation field, some flux may also be emitted in molecules like CO or HCN, in particular in the low-lying rotational transitions, and there are further fine-structure lines that may contribute as well. As mentioned in the introduction, there are plenty of observations that studied emission from starburst galaxies with present-day sub-mm telescopes [e.g. Omont et al., 1996; Carilli et al., 2002; Walter et al., 2004; Klamer et al., 2005; Weiß et al., 2005; Maiolino et al., 2007; Walter et al., 2007; Riechers et al., 2008b,a; Greve et al., 2009; Riechers et al., 2009; Walter et al., 2009a].

With the upcoming sensitivity of ALMA, we expect that such PDRs can be probed in more detail as well. Therefore, bright lines like the [CII] 158 μm line can be detected at higher significance, allowing higher spectral resolution and probing the velocity structure of

7.3 Observables in the PDR

Observable	λ [μm]	φ [mJy]	min. Redshift
[O I] $^3P_1 \to {}^3P_2$	63.2	~ 0.18	$z > 5.7$
[O III] $^3P_1 \to {}^3P_0$	51.8	~ 0.18	$z > 3.8$
[N II] $^3P_2 \to {}^3P_1$	121.9	~ 0.07	$z > 2.5$
[O I] $^3P_0 \to {}^3P_1$	145.5	~ 0.03	$z > 1.9$
[C II] $^2P_{3/2} \to {}^2P_{1/2}$	157.7	~ 0.7	$z > 1.7$
[S III] $^3P_1 \to {}^3P_0$	33.5	~ 1.5	$z > 11.1$
[Si II] $^2P_{3/2} \to {}^2P_{1/2}$	34.8	~ 0.06	$z > 11.1$

Table 7.1: The main observables for ALMA in PDRs at high redshift. This specific example assumes a galaxy with a starburst as in NGC 1068 placed at $z = 8$. We also give the minimal redshift from which the lines would be redshifted into the ALMA bands.

the gas in more detail. Also, weaker lines may be detected as well, probing gas at different densities and providing additional information on the chemical conditions.

band	freq. [GHz]	θ_{res} ["]	S_c [mJy]	S_l [mJy]	θ_{beam} ["]
3	84 – 116	0.034	0.019	0.163	56
4	125 – 169	0.023	0.023	0.174	48
5	163 – 211	0.018	0.298	2.63	35
6	211 – 275	0.014	0.039	0.225	27
7	275 – 373	0.011	0.077	0.372	18
8	385 – 500	0.008	0.143	0.620	12
9	602 – 720	0.005	0.232	0.813	9

Table 7.2: Frequency range, angular resolution θ_{res} at the largest baseline, line sensitivity S_l for a linewidth of 300 km/s and continuum sensitivity S_c for 3σ detection in one hour of integration time and primary beam size θ_{beam}. 3 more bands might be added in the future, band 1 around 40 GHz, band 2 around 80 GHz and band 10 around 920 GHz, which will have similar properties as the neighbouring bands.

To obtain a rough estimate on the expected PDR fluxes in different lines, we have evaluated the PDR fluxes that we would expect for a system like the Seyfert 2 galaxy NGC 1068, if it were placed at high redshift. We adopt $z = 8$. This system consists of a central X-ray dominated region [Galliano et al., 2003] and a circumnuclear starburst ring of ~ 3 kpc in size, with a stellar mass of ~ 10^6 $M_\odot$ and an age of 5 Myr [Spinoglio et al., 2005]. On scales of a few hundred parsecs, one finds a star formation rate of a few times 10 $M_\odot$ yr^{-1} kpc^{-2} [Davies et al., 2007]. This is close to the star formation rate in Eddington-limited starbursts as suggested by Thompson et al. [2005]. To understand which of the fluxes emitted in this

7. PROBING HIGH-REDSHIFT QUASARS WITH ALMA[1]

region would be detectable with ALMA if this system were located at $z = 8$, we went through the spectroscopic sample of Spinoglio et al. [2005] and checked which lines would be redshifted into the ALMA frequency bands, and what would be the expected amount of flux. The fluxes given by Spinoglio et al. [2005] have been measured with a frequency resolution of 1500 km/s. Correspondingly high velocities can indeed be reached in the presence of fast jets or outflows. However, typical line profiles show that most of the flux is in a range of ±150 km/s, which we adopt here as a fiducial value. The so obtained observable line transitions are summarized in Table 7.1, while the expected sensitivity and angular resolution at the largest baselines is given in Table 7.2. For all the transitions, a 3σ detection seems possible for an integration time of a few hours. The table does not include CO transitions, as the PDR would only excite the low-lying transitions which would not fall in ALMA's frequency range for $z > 8$. We note that the expected ratio between [NII] and [CII] is comparable to the observational upper limit derived by Walter et al. [2009b].

7.4 Expectations for the XDR

As ALMA may for the first time detect and resolve emission for the central X-ray dominated regions, we want to assess here in more detail the expected chemical conditions in the central region and the corresponding fluxes in different lines. We start by discussing the implications of X-rays for the conditions in molecular clouds. We then show how ALMA observations can distinguish between X-ray chemistry and an intense burst of star formation on the same spatial scales. Afterwards, we provide a set of systematic model predictions, first assuming an XDR of constant size, but also considering the potential increase of the XDR in case of a higher X-ray luminosity.

7.4.1 Implications of X-rays for molecular clouds

We follow the chemistry in a one-dimensional molecular cloud complex irradiated by X-rays with the XDR code of Meijerink & Spaans [2005]. The model includes more than 50 chemical species and several thousand reactions. For CO, the detailed level populations are solved consistently with the 1D radiation transport equation [Poelman & Spaans, 2005, 2006]. As the low-metallicity case was explored in detail by Spaans & Meijerink [2008], we focus here in particular on situations with about solar metallicity.

The first model we discuss corresponds to the XDR in the Seyfert 2 galaxy NGC 1068 (see § 7.5.1). We plot chemical abundances and CO emission for cloud column densities between $10^{20} - 10^{24}$ cm^{-2} in Fig. 7.3. Larger column densities correspond to extreme ULIRGs like Arp 220 that is even optically thick around 350 GHz (P. Papadopoulos, private communication). The fiducial gas density of 10^5 cm^{-3} has little impact on our results, unless it

7.4 Expectations for the XDR

drops to below $10^{4.5}$ cm^{-3}. Above this limit, the XDR properties are determined by the ratio of X-ray flux to gas density. For lower densities, emission in the high-J CO lines would not be excited due to the critical densities. However, high-density gas appears to exist in the center of NGC 1068 [Galliano et al., 2003].

The strong X-ray flux of $\sim$ 170 erg s^{-1} cm^{-2} in NGC 1068 suffices to make the gas essentially atomic and leads to high temperatures of $\sim$ 3000 K, as well as relatively low CO abundances of the order 10^{-7} due to photodissociation by soft UV-photons produced after the absorption of X-rays. However, the CO abundance is still higher than in typical PDRs, and the CO intensity is high, due to the strong thermal excitation in the hot gas. For a column of 10^{22} cm^{-2}, our results appear of the same magnitude as in the model of Galliano et al. [2003]. For larger columns, the temperature gradually decreases, the gas becomes molecular and CO gets more abundant, and we find intensities of the order 10^{-2} erg s^{-1} cm^{-2} sr^{-1} in the high-J CO lines.

To explore the dependence of the chemistry on the X-ray flux, we consider two additional cases. An extreme case with $\sim$ 1 erg s^{-1} cm^{-2} is shown in Fig. 7.4. In this model, we find lower temperatures of $\sim$ 70 K, a large fraction of molecular gas and CO abundances of the order of 10^{-4}. While the lower temperature tends to decrease the CO line intensities, they are still enhanced due to the larger CO abundance. Above a column of 10^{23} cm^{-2}, the intensities increase rather slowly as the lines become optically thick.

As an intermediate scenario, we consider a source with an X-ray flux of $\sim$ 10 erg s^{-1} cm^{-2} (see Fig. 7.5). In this model, the temperature is increased to $\sim$ 100 K. The CO abundance is initially of the order 3×10^{-6} and increases to $\sim 10^{-4}$ for larger columns. For columns less than 10^{22} cm^{-2}, the intensities are thus reduced by about an order of magnitude compared to the previous case, while they are increased by an order of magnitude for larger columns.

7.4.2 Separating the XDR from a nuclear starburst

In the center of an active galaxy, not only the X-ray emission is enhanced, but one may expect the presence of a strong nuclear starburst. For instance, Arp 220 harbors such a starburst on scales of $\sim$ 300 pc. We therefore compare the expected CO line SED of a strong starburst with $G_0 = 10^5$ with the CO line SEDs in X-ray dominated regions, based on the models provided by Meijerink et al. [2007]. We normalize them such that the CO (10-9) transition has the same intensity in all models. In this case, the SEDs can hardly be distinguished at the low-J transitions that are typically observed at low redshift (see Fig. 7.6). At higher-J transitions, the PDR SED drops considerably and flattens on a low level due to the small amount of hot gas in the outer layer of the molecular cloud. We expect that a value of $G_0 = 10^5$ is a robust upper limit for the soft-UV flux that can be obtained in a galaxy. In fact, larger values have never been indicated in previous observations, and indeed such a value would require extreme conditions as in the Orion Bar throughout all of the galaxy. In XDRs,

7. PROBING HIGH-REDSHIFT QUASARS WITH ALMA

a much larger fraction of the gas is at high temperatures, and thus the SED is not expected to drop as rapidly.

If the total amount of energy injected by X-rays and by soft-UV photons is comparable (within a factor of 10), then the presence of X-rays can be clearly inferred using the CO (16-15) transition, if the local X-ray flux is at least 2.8 erg s^{-1} cm^{-2}. However, as discussed in § 7.2, we expect X-ray flux to dominate over the soft-UV in the center of the galaxy. Observations at even higher-J transitions may be useful to determine the local amount of X-ray flux from the CO line SED. A potential uncertainty is the presence of cold dust, which may to some degree absorb the CO line emission and thus change the appearance of the SED. Due to the characteristic scaling of dust absorption with wavelength, we expect that such a behavior could be recognized and potentially corrected. For this purpose, it is of course desirable to measure as many high-J CO lines as possible.

7.4.3 Model predictions for XDRs of constant size

Although we have shown in § 7.2 that the central XDRs can likely be resolved with ALMA, it is currently not clear how the expected XDR size varies with X-ray luminosity. If the strengths of the soft-UV field is independent of this, one should on average expect a larger XDR for higher X-ray luminosities. However, it is also conceivable that the X-ray luminosity is indicative of the system as a whole, and that a higher X-ray luminosity may be accompanied by a stronger soft-UV field. In such a case, one might expect a smaller increase in the XDR size or even a constant size. For this reason, we will consider two extreme cases, assuming that more realistic scenarios should lie in between the two. In this subsection, we will assume that the size of the XDR is always constant, of ~ 200 pc. Of course, the numbers given here can be easily rescaled for other XDR sizes, or for area-filling factors smaller than 1. In the following subsection, we will then discuss the implications of varying the X-ray luminosity for constant G_0.

For a first estimate, let us consider at source at $z = 5$ with an XDR of at least 100 pc, corresponding to an angular scale of 0.016", and a typical intensity in the high-J CO lines of 10^{-3} erg s^{-1} cm^{-2} sr^{-1}. As can be seen from the calculations above, such an intensity can be reached in a broad range of systems for column densities of at least 10^{23} cm^{-2}, and in fact also for column densities of at least 10^{22} cm^{-2} in the presence of sufficient X-ray flux. With a fiducial velocity dispersion of 300 km/s, this corresponds to a flux of 0.03 mJy, which is detectable in a bit more than a day in ALMA band 6.

In a similar way, it is possible to calculate the expected fluxes also in various fine-structure lines. Based on the detailed parameter study provided by Meijerink et al. [2007][1], which shows the expected fluxes in various lines as a function of X-ray luminosity density and column density, we therefore provide detailed predictions for the fluxes from the central

---

[1] http://www.strw.leidenuniv.nl/meijerin/grid/

7.4 Expectations for the XDR

X-ray dominated region, assuming a characteristic size of 200 pc, consistent with the results obtained in § 7.2.

For the high-J CO lines, we focus on those which are redshifted into ALMA band 6, which offers a good compromise between angular resolution (0.014" at the largest baseline) and sensitivity (~ 0.04 mJy for an integration time of one day, 3σ detection and a line width of 300 km/s). At a redshift $z = 5$, this corresponds to the (10-9) CO transition, at $z = 8$ to the (14-13) CO transition. We assume that the projected surface area is homogeneously filled with molecular clouds with central densities of 10^5 cm^{-3}. Fig. 7.7 shows how the expected flux varies as a function of the cloud column density and the X-ray luminosity[1]. Similar results would be obtained for cloud densities of $10^{4.5}$ cm^{-3}, while it is more difficult to excite CO emission at lower densities.

The figure illustrates that higher fluxes can be obtained for larger column densities, while intermediate X-ray fluxes are ideal for stimulating emission in the CO lines considered here. This is because at very high fluxes, a significant amount of X-rays would be converted into soft UV-photons and dissociate the molecules.

In Figs. 7.8 and 7.9, we show the corresponding results for the [CII] 158 μm line, both for a density of 10^5 cm^{-3} and a density of 10^4 cm^{-3}. For densities of 10^5 cm^{-3}, low column densities are sufficient to yield a detectable amount of flux, and the flux monotonically increases with the X-ray luminosity. As in the PDR case, this line is therefore valuable to explore the centers of high-redshift quasars, and provides complementary information to the CO lines. For densities of 10^4 cm^{-3}, we find a stronger dependence on column density, and indeed columns of at least 10^{23} cm^{-2} are needed to yield a detectable amount of flux.

The results for the [OI] 63 μm line are given in Figs. 7.10 and 7.11, again for densities of 10^5 cm^{-3} and 10^4 cm^{-3}, respectively. This line quickly becomes optically thick. Therefore, in particular for cloud densities of 10^5 cm^{-3}, it is insensitive to the column density, but provides a good measure for the X-ray flux. It is very bright. Even for cloud densities of 10^4 cm^{-3}, it depends more on X-ray luminosity than column density, and is still detectable in the case of high column densities and strong X-ray fluxes. It is thus well-suited to study gas dynamics in the central XDR by resolving the line profile.

We also provide results for the [OI] 146 μm line in Fig. 7.12, for a cloud density of 10^5 cm^{-3}. For lower densities, this line is hard to excite. It can also be very bright and shows a strong dependence on the X-ray flux. The ratio between the [OI] 63 μm line and the 145 μm line is generally about 0.1.

Emission from neutral carbon seems more difficult to detect. The intensity of the [CI] 369 μm line is typically at least an order of magnitude smaller than the intensity in the [CII] 158 μm line, and only significant in the case of strong X-ray fluxes and high column densities. As discussed in § 7.5.3, the relatively low ratio between these lines in the recently detected

---

[1] For the conversion between X-ray luminosity and flux, we assumed optically thin conditions. One may need to correct for further attenuation in case of large filling-factors.

7. PROBING HIGH-REDSHIFT QUASARS WITH ALMA[1]

$z = 6.42$ quasar puts a significant constraint on theoretical models. Other carbon lines like [CI] 609 μm or [CI] 230 μm are even weaker and should never be visible. Additional lines like [SiII] 35 μm or [FeII] 26 μm can be bright as well [Meijerink et al., 2007], but typically have too short wavelengths for ALMA, except at redshifts $z > 9$.

The reader may notice that for the model predictions given in this subsection, we restricted ourselves to a range of X-ray luminosities of about two orders of magnitude, corresponding to the range of X-ray fluxes avaiable in the data by Meijerink et al. [2007]. This range of data has been chosen such that it covers the observationally interesting cases. For lower fluxes, or in our case lower X-ray luminosities, we would not expect significant emission driven by X-rays, as can be seen in the corresponding figures. Similarly, it is straightforward to extrapolate the behavior towards higher X-ray fluxes: As shown in Fig. 7.7, the CO intensities decrease considerably at higher luminosities. This is because part of the X-rays are reprocessed to soft-UV photons that dissociate all the CO. At even higher intensities, the gas temperature would increase up to 10^4 K and become fully ionized. Molecules would no longer survive under such circumstances. The emission of [CII] may increase a bit further, until a large fraction of carbon is doubly-ionized at temperatures near 10^4 K. Similar considerations hold for the oxygen line. However, it is not clear whether such scenarios are actually physical, or if the molecular cloud would rather be photo-evaporated at that point. As already mentioned above, the size of the XDR may be more extended in the case of such high luminosities. Then, moderate X-ray fluxes will be present on larger scales.

In summary, we can therefore conclude that the main observables for the central XDR are the high-J CO lines as well as the fine-structure lines of [CII] and [OI]. The [OI] 63 and 146 μm lines show a strong dependence on X-ray luminosity and may provide a good handle on this quantity. In combination with an observation of the [CII] or high-J CO lines, the X-ray luminosity can be estimated as well. As shown by Meijerink et al. [2007], such line ratios also provide valuable information on gas density and temperature. We also note that it is possible to discriminate such XDRs from regions with strong mechanical heating [Papadopoulos et al., 2008]. This can be done for instance on the basis of the observed X-ray luminosity, or by looking at the dust SED. While in XDRs, both dust and gas will be at high temperatures in dense clouds, there should be a clear discrepancy between gas and dust temperature if heating is due to local shocks. We therefore expect that the physical conditions in the central XDRs can be probed in detail with ALMA.

7.4.4 Model predictions for variable XDR-sizes

As mentioned above, the size of the XDR may increase considerably for increasing X-ray luminosity. We first explore the dependence on the average density and the X-ray luminosity, assuming a typical cloud column density of 10^{23} cm^{-2}, and a soft-UV field $G_0 = 100$. The X-ray spectrum is assumed to range from 1 to 100 keV. As shown in § 7.2, the size of the

XDR then varies as a power-law with density. The same is true for the expected fluxes of CO, [CII] and [OI], as shown in Fig. 7.13. We estimate those by adopting a typical value in mJy on scales corresponding to the size of the XDR. These should indeed consitute the main contribution. For [CII] and [OI], the total flux may be a bit larger, as the expected amount of emission increases with the X-ray flux (see previous subsection). CO, on the other hand, might be efficiently dissociated in the very inner core. However, we do not expect this to affect our predictions significantly.

We also explore the role of the molecular cloud column density for the expected fluxes. For this purpose, we adopt an average density of 10^4 cm^{-3} and a soft-UV field $G_0 = 100$. In Fig. 7.14, we show the expected fluxes as a function of X-ray luminosity and column density. As a generic feature, we find that the expected fluxes vary for columns smaller than 10^{23} cm^{-2}, but level-off for higher values. The fluxes in high-column density systems should thus essentially depend on the X-ray luminosity only.

7.5 Evidence for XDRs and the interpretation of sub-mm line observations

Although the presence of a central X-ray dominated region seems unavoidable from a theoretical point of view, we want to review current evidence for the presence of central XDRs in active galaxies. Such evidence is present in local galaxies like NGC 1068 that can be studied in great detail, as well as in high-redshift quasars like APM 08279 in which the corresponding fluxes are magnified by a gravitational lens. Similar indications are present also in the Cloverleaf quasar, where CO fluxes up to the (9-8) transition have been measured, and no turnover in the line SED has been found yet [Bradford et al., 2009]. These sources will allow a first test for the expectations we have formulated in the previous section once the ALMA telescope becomes available. We also discuss the $z = 6.42$ quasar SDSS J114816.64+525150.3, as it provides a good and interesting example concerning the complex interpretation of sub-mm line observations. We will further discuss how its properties can be understood better if additional data are provided, with particular focus on the role of ALMA.

To distinguish between different excitation mechanisms of molecular clouds, it is very important to have observational diagnostics for the various excitation mechanisms. First efforts for modelling the chemistry in XDRs were performed by Maloney et al. [1996] and Lepp & Dalgarno [1996], while PDR chemistry was originally studied by Hollenbach & Tielens [1999]. Recent efforts to discriminate such models have been performed by Pérez-Beaupuits et al. [2007, 2009], and new diagnostic diagrams to discriminate AGN- and starburst-dominated galaxies have been provided by Spoon et al. [2007] and Hao et al. [2009].

7. PROBING HIGH-REDSHIFT QUASARS WITH ALMA[1]

7.5.1 NGC 1068

As shown by Galliano et al. [2003], NGC 1068 contains a central XDR. Their analysis is based on the observed intensitiy of the H_2 2.12 μm line [Galliano & Alloin, 2002] and the rotational CO lines [Schinnerer et al., 2000], and showed that the observed emission can be consistently explained with the XDR-model of Maloney et al. [1996] under the following assumptions:

- The central engine is a power-law X-ray source with spectral slope $\alpha = -0.7$ and luminosity of 10^{44} erg s^{-1} in the 1 – 100 keV range, consistent with the X-ray luminosity determined from VLBI water maser observations [Greenhill et al., 1996].

- The emission originates from molecular clouds with a density of 10^5 cm^{-3}, a column of 10^{22} cm^{-2} at a distance of 70 pc and solar metallicity.

They show that the central XDR is indeed inhomogeneous due to a central X-ray absorber that shields the X-rays along the line of sight, while they can stimulate molecular emission in the perpendicular direction. A sketch of such a situation is given in Fig. 7.15. In the presence of a torus, indeed it seems likely that X-ray emission is preferred in the direction orthogonal to the torus, and the XDR fluxes may therefore be preferentially detectable in situations where the molecular disk is observed face-on.

7.5.2 APM 08279

In the $z = 3.9$ galaxy APM 08279, the usual geometric effects concerning emission from the central regions are compensated by a gravitational lens. It therefore provides an ideal test case to study the emission we might see when future telescopes like ALMA can observe the central regions with higher sensitivity and higher spatial resolution. As expected for molecular clouds excited by X-rays, they find that the CO line intensity increases up to the (10-9) transition. Significant flux is also detected in the HCN(5-4) line.

Based on brightness temperature arguments, the results from high-resolution mapping and lens models from the literature, Weiß et al. [2007] show that the molecular lines arise from a compact (100 – 300 pc), highly gravitationally magnified (m=60-110) region surrounding the central AGN. It is interesting to note that this amount of magnification is comparable to the increase in sensitivity due to ALMA. They can distinguish two components of the gas: A cool component with a density of 10^5 cm^{-3} and a temperature of 65 K, on scales larger than 100 pc, and a warm component of gas with 10^4 cm^{-3} and temperatures of 220 K on scales smaller than 100 pc. So far, their results do not provide indications for inhomogeneities.

We have compared the turnover for the rotational CO lines found by Weiß et al. [2007] with the grid of PDR and XDR model predictions that were made publicly available by

7.5 Evidence for XDRs and the interpretation of sub-mm line observations

Meijerink et al. [2007]. In the PDR case, we find that cloud densities of 10^5 cm^{-3} and a radiation field of $G_0 \sim 10^{4.75}$ with a cloud column of $\sim 6 \times 10^{22}$ cm^{-2} are required. For the XDR case, the turn-over can be explained with a cloud density of $10^{4.25}$ cm^{-3}, an X-ray flux of 2.8 erg s^{-1} cm^{-2} and a cloud column density of $\sim 2.6 \times 10^{22}$ cm. In both cases, we need to require that additional cold gas is present that emits in particular in the low-lying J levels to explain the observed ratio between high-J and low-J rotational CO lines.

The X-ray spectrum of APM 08279 has been observed with Chandra [Chartas et al., 2002], indicating a luminosity of $4 \times 10^{46} m^{-1}$ erg s^{-1}. Adopting a lens magnification of $m \sim 100$, one expects a flux of ~ 354 erg s^{-1} cm^{-2} at a distance of 100 pc for an optically thin source. However, for a density of $10^{4.25}$ cm^{-3}, we expect significant attenuation effects that may considerably decrease the flux in the cloud. As shown by Wada et al. [2009], shielding may locally vary by two orders of magnitude due to column density fluctuations in the torus and give rise to a peak optical depth of $\tau \sim 5$ for frequencies of 3 keV.

With ALMA, it will be possible to probe the distribution of these gas components and their kinematics in even more detail. Due to the higher sensitivity, the error bars in the flux measurement will decrease further and one may probe whether the turnover occurs at the (10-9) transition, as indicated now, or indeed even at higher-J levels. In this case, one could clearly discriminate between PDR and XDR models.

7.5.3 SDSS J114816.64+525150.3

The discovery of the extended source SDSS J114816.64+525150.3 at $z = 6.42$ by Walter et al. [2009a] has stimulated great observational interest that led to a broad collection of data on this source. Recently, Riechers et al. [2009] provided a table with fluxes in the CO (3-2), (6-5) and (7-6) transitions, fluxes in the CI line and the [CII] line, as well as upper limits on five additional lines. Although the publication of such upper limits is very valuable and can provide important constraints on theoretical models, for simplicity we focus here on those lines that were actually detected. As we shall see, the interpretation of those is already complex, and a more thorough analysis would be beyond the scope of this work. The models and data empoyed here are based on the publicly available grids provided by Meijerink et al. [2007]. Due to the large scales of this source, we assume that most of the flux is driven by PDR chemistry, especially because the relevant scales for XDRs are currently unresolved.

To understand the complexity of these data, it is illustrative to check whether a model with one given density and a fixed parameter G_0 is able to explain them. For CO, the ratio between the (6-5) and the (3-2) transition is about 3.35, while the ratio between (7-6) and (3-2) is about 3.15. Although it appears that the turnover in the SED may have been reached, this is not fully clear from the data, as the uncertainty in the fluxes is $\sim 10\%$. The increased intensity in the (6-5) transition requires a significant amount of soft-UV flux, while the almost flat behavior indicates that the gas should be at intermediate densities of about

7. PROBING HIGH-REDSHIFT QUASARS WITH ALMA[1]

$\sim 10^4$ cm^{-3}. At higher densities, the seventh rotational level would be more easily excited, which would give rise to a significant discrepancy between the (6-5) and the (7-6) transition, even for low values of G_0. For lower densities, on the other hand, it becomes difficult to excite these rotational levels. A reasonable fit to the CO SED can thus be obtained with $n = 10^4$ cm^{-3} and $G_0 = 10^{1.75}$.

A problem arises, though, if the observed fluxes in [CII] and CI should arise from the same gas component. In this case, our model predicts a flux ratio of [CII] to CI of ~ 600, and the intensity of [CII] and CI would be much higher than the CO intensities. We note, however, that the fine-structure lines are already optically thick at these densities, while the CO lines are optically thin. Thus, additional clouds may be present that enhance the CO intensities, while the optically thick emission in [CII] and CI is unaffected. Nevertheless, as observations show a corresponding line ratio of 17.7, these lines should originate from a different gas component. Our models require the presence of additional gas clouds with density of $\sim 10^3$ cm^{-3}, $G_0 \sim 10^{1.75}$ and large columns $N_H > 10^{21}$ cm^{-2}, at which enough of the soft-UV flux is shielded to yield a low ratio between [CII] and CI. This line ratio is indeed strikingly low, as PDR models generally predict larger numbers for this ratio, and also in this case the match is not perfect, but only within the 20% error of the CI measurement. Although this may be an issue due to uncertain abundance ratios, a more precise determination of the flux in this line is highly desirable.

So far, we have postulated two gas components to explain the CO line SED and the fluxes from [CII] and CI. It is however important to cross-check whether these components may affect the line ratios that the other component should reproduce. For the CO line SED, indeed the intensity in a low-density cloud is smaller by one order of magnitude. The high-density component, however, gives rise to an intensity in [CII] that is smaller than the corresponding intensity from the other component by just a factor of 5. To avoid that this perturbs the [CII] to CI flux ratio, we need to require that the low-density gas is more abundant than the gas at high densities. In this case, it seems however likely that the low-density gas would perturb the CO line SED significantly.

The most probable explanation in terms of a two-component picture is thus that indeed the low-density component is more abundant and explains the observed fine-structure lines. The impinging soft-UV flux must be moderate in order to reconcile the low ratio between these lines. To still explain the observed CO line SED, we need to require that the high-density component is at higher temperatures, due to an increased value of $G_0 \sim 10^3$. As one generally expects that high-density gas is exposed to a weaker radiation field due to shielding effects, this component should be spatially separated in regions of strong active star formation. Alternative scenarios are however feasible as well. For instance, the observed [CII]/CI ratio may also be produced in the presence of a weak X-ray background rather than a soft-UV field, or if a the cosmic-ray background is enhanced by a factor of 10-100 compared to the Milky Way [see also Meijerink et al., 2006].

Of course, this model is still an oversimplification, as one may indeed expect a variety of different densities in both star-forming and more quiescent parts of the galaxy. In addition, there are theoretical uncertainties concerning the metallicity and the abundance ratios that may affect our results. Nevertheless, this model reproduces the main features in the observed fluxes and illustrates the difficulties in simultaneously modeling observations in different lines. This challenge will become more severe, as the high sensitivity of ALMA will allow us to study even more lines. On the other hand, it is certainly desirable if some of the simplest models can be discarded on such grounds. We note that such models are also predictive. For the scenario described here, we would for instance expect that also the flux in the [OI] 63 μm line and the [OI] 145 μm line originates predominantly from the low-density gas component. In this case, the ratio of these fluxes to the flux in [CII] would be $0.06 - 0.1$ and 8×10^{-4}, respectively. If the oxygen abundance is enhanced by a factor of 4 compared to the galactic abundance, the flux in the 63 μm line would be larger by a factor of 2.

With future ALMA measurements, we expect that the CO line SED can be probed in more detail and with higher accuracy at higher-J levels, to check whether the turnover has already been reached. Apart from the observed gas component that indicates a turnover near the $J = 6$ transition, a higher density component may be present that can more easily excite flux at higher-J levels. The increased sensitivity and angular resolution of ALMA may help to detect spatial variations in the different fluxes, which may help to discriminate regions of intense star formation from more quiescent zones. In the center of the galaxy, ALMA can check for the presence of an X-ray dominated region. Significant progress may however be possible in the mean time, for instance by measuring additional high-J CO lines or by a more accurate determination of the [CII] to CI flux ratio.

7.6 The expected number of sources

As shown in § 7.3, large-scale PDR fluxes, for instance in the [CII] 158 μm line or the [OI] 146 μm line, are sufficiently large for detection in a few hours. Fluxes from the XDR originate from a smaller region, but may also give rise to a significant flux component. With an integration time of a few hours, active galaxies with a comparable brightness as NGC 1068 should be detectable in an ALMA field of view. One might therefore wonder whether a search for active galaxies, based on the before-mentioned fine-structure lines, is feasible. Beyond that, it is of strong interest to know whether there are reasonable chances to find active galaxies with other surveys such as JWST. In such cases, ALMA could perform relevant follow-up studies that probe the central X-ray dominated regions and the gas dynamics within them. If a sufficient number of sources is obtained, ALMA can probe the feeding of black holes between redshift 6 and 10 and thus constrain models concerning the growth of

7. PROBING HIGH-REDSHIFT QUASARS WITH ALMA[1]

supermassive black holes based on the diagnostics provided here. We therefore conclude this paper with an estimate on the high-redshift black hole population.

To assess this possibility in more detail, we start with some general considerations regarding the black hole population at $z > 6$. Then we discuss estimates on the number of sources and the possibility to detect them with an ALMA deep field.

7.6.1 Black hole growth at high redshift

As shown recently by Shankar et al. [2009], a supermassive black hole with 10^9 $M_\odot$ today had, on average, 10^7 $M_\odot$ at $z = 5$. Therefore, one needs to explain how supermassive black holes have accreted this mass at early times. Indeed, some of them need to accrete even faster, as 10^9 $M_\odot$ black holes have already been detected at $z > 6$ [e.g. Fan et al., 2006]. We will however focus on the more conservative case of 10^7 $M_\odot$ black holes.

We follow Shapiro [2005] and describe black hole growth by accretion using the formula,

$$\frac{dM_{\rm BH}}{dt} = \frac{M_{\rm BH}}{\tau_{\rm growth}}, \tag{7.2}$$

where

$$\begin{aligned}\tau_{\rm growth} &= 0.0394 \frac{(\epsilon_M/0.1)}{1-\epsilon_M} \frac{1}{\lambda_E} \frac{1}{f_{\rm duty}} \text{ Gyr}\\ &\equiv 0.0394 p \text{ Gyr},\end{aligned} \tag{7.3}$$

and where $\epsilon_M = L/\dot{M}_0 c^2$ is the efficiency of conversion of rest-mass energy to luminous energy, λ_E the Eddington ratio and $f_{\rm duty}$ the duty cycle, i. e. the fraction of quasars active at a time. The parameter p summarizes the complicated dependence on the latter parameters. With the parameters inferred by Shankar et al. [2009] for the observed high-redshift black hole population up to $z = 6$, we obtain $p \sim 25$.

The dependence of the black hole mass evolution in this parameter is given in Fig. 7.16. For $p > 7$, the black hole mass hardly evolves with redshift, implying that black holes would need to form with extremely high masses. There are, however, no theoretical models available that would explain such massive seeds, which would have comparable masses to the first galaxies at high redshift.

It is more likely that the black hole population evolved with redshift, implying that a larger fraction of them was active and had higher Eddington ratios. For instance, a value of $p = 2.5$ requires intermediate mass black holes with $10^4 - 10^5$ $M_\odot$ to be the progenitors of the first supermassive black holes. Such scenarios have been suggested by Eisenstein & Loeb [1995], Koushiappas et al. [2004], Begelman et al. [2006], Spaans & Silk [2006], Lodato

7.6 The expected number of sources

& Natarajan [2006] and Dijkstra et al. [2008]. For even smaller values of p, supermassive black holes could even originate from stellar progenitors.

In this respect, it is speculated that so-called dark stars, which would be stars powered by dark matter annihilation rather than nuclear fusion [Spolyar et al., 2008; Freese et al., 2008a], could obtain even larger masses than conventional Pop. III stars. However, the high masses suggested in these works are not confirmed by other authors [see also work by Iocco, 2008; Iocco et al., 2008] and may be in conflict with the observed reionization optical depth [Schleicher et al., 2008b,a].

In all of these cases, it is evident that the black hole would need to grow considerably by accretion, and that a change in the parameters describing the black hole population is required. In particular, a larger fraction of active quasars and a higher Eddington efficiency are required to obtain the black hole masses that we observe today. As pointed out by Kawakatu & Wada [2009], even super-Eddington accretion may be required to obtain the observed black hole masses.

7.6.2 Estimates on the number of sources

To estimate the expected number of high-redshift black holes, we will show extrapolations of theoretical models describing the observed high-redshift black hole population, as well as estimates based on the number of $10^9 \, M_\odot$ black holes in the local universe.

To translate a given black hole number density into the number of sources in a certain field of view, we adopt a similar formalism as Choudhury & Ferrara [2007] and count the number of black holes $N(z, M_{BH})$ with masses larger than M_{BH} in a redshift interval $[z, z + \Delta z]$ per solid angle. This is given as

$$\begin{aligned} N(z, M_{BH}) &= \int_z^{z+\Delta z} dz' \frac{dV}{dz' d\Omega} \int_{M_{BH}}^{M_{max}} d\log M'_{BH} \\ &\times f_a \frac{dn}{d\log M_{BH}}, \end{aligned} \quad (7.4)$$

where $dV \, dz'^{-1} \, d\Omega^{-1}$ denotes the comoving volume element per unit redshift per unit solid angle, which is given as [Peebles, 1993]

$$\frac{dV}{dz' d\Omega} = D_A^2 c \frac{dt}{dz}. \quad (7.5)$$

In this expression, D_A is the angular diameter distance, c is the speed of light and $dt/dz = 1/(H(z)(1+z))$, where $H(z)$ is the expansion rate as a function of redshift. The term $dn/d\log M_h$ describes the number density of dark matter halos per unit mass, and the maximum black hole mass is taken as $M_{max} = 10^9 \, M_\odot$.

7. PROBING HIGH-REDSHIFT QUASARS WITH ALMA[1]

As explained above, we focus here on the possibility to search for fine-structure line emission. With a bandwith of 8 GHz, we expect that ALMA could scan a redshift interval of $\delta z = 0.2$ at $z \sim 8$ in bands 6 and 7 within an integration time of a few hours. We assume here that this process is iterated until a total redshift interval of $\Delta z = 0.5$ is covered, which we adopt as a fiducial redshift range in our calculation. Alternatively, one could focus on the detection of continuum flux, which would allow to scan an even larger redshift interval within a given time, but at the same time it would be more difficult to infer the redshift of source.

For our first estimate that extrapolates theoretical models of Shankar et al. [2009] and Treister et al. [2009], we use their results for quasars with luminosities larger than 10^{44} erg s^{-1}, which can be associated with a black hole mass of 10^7 $M_\odot$ for an Eddington ratio of 10%, as they find in their calculations. Note that Treister et al. [2009] only provide numbers for Compton-thick quasars, but give a Compton-thick quasar fraction of $\sim$ 10%, which we use to infer the total population. The results based on this approach are given in Fig. 7.17 and are only approximate.

The other approach is based on the number of 10^9 $M_\odot$ black holes in the local universe. As noted above, such black holes had an average mass of 10^7 $M_\odot$ at $z = 5$ [Shankar et al., 2009]. Therefore, the comoving number density of such black holes can be estimated using the number density of today's 10^9 $M_\odot$ black holes, which is $\sim 10^{-3.5}$ Mpc^{-3} [Shankar et al., 2009]. It may be even higher, as Compton-thick black holes at $z = 0$ are hard to detect [Treister et al., 2009]. The expected number of sources based on this estimate is shown in Fig. 7.17.

One may note that the number of sources obtained in this way is larger than the value from the extrapolation of theoretical models, i. e. by Shankar et al. [2009] and Treister et al. [2009]. Such a behavior can be expected from our discussion in § 7.6.1: As supermassive black holes should have higher duty cycles at $z > 6$, the models by Shankar et al. [2009] and Treister et al. [2009] will naturally underestimate this population. At the same time, the estimate based on the present-day 10^9 $M_\odot$ population is also an upper limit, since not all of these black holes will be active at the same time. We therefore expect that these approaches yield an upper and a lower limit and that the actual black hole population will lie in between.

For the expected abundance of 10^7 $M_\odot$ black holes at $z = 5$, we therefore adopt a fiducial number of ~ 0.08 in a solid angle of $(1')^2$, and a number of ~ 0.03 at $z = 8$. We estimate the uncertainty to be half an order of magnitude. Based on the large fluxes found in § 7.5.1 and the results of Spaans & Meijerink [2008], we argue that even smaller systems with 10^6 $M_\odot$ black holes should be detectable. Calculations by Shankar et al. [2009] show that the cumulative space density of high-redshift quasars increases by almost an order of magnitude if the minimum luminosity is lowered by an order of magnitude. Additional sources may be present as well, such as starburst galaxies or ULIRGS. One may therefore expect one active galaxy in a field of view of one arcmin^2 at $z \sim 6$.

At higher redshift, this number may however decrease significantly. Therefore, searching for new sources in an ALMA deep field seems difficult for $z > 6$, and it is important that surveys as performed by JWST[1] provide a catalogue of high-redshift sources. Based on the estimates discussed here, we conclude that this may be sufficient to study black hole accretion between redshifts 6 and 10, and perhaps even beyond.

7.7 Conclusions

In this paper, we have explored how ALMA observations may help to extend our current knowledge concerning the observations of high-redshift quasars. An obvious result is that current observations of the starburst component can be done with higher resolution and higher sensitivity, which may therefore provide detailed local gas kinematics. Due to the higher sensitivity, it seems likely that also a number of the weaker fine-structure lines may be detected, like [NII] 122 μm for $z > 2.5$, or [SIII] 34 μm and [SiII] 35 μm for $z > 11$, providing a valuable independent probe of the gas chemistry.

Due to the combination of high resolution and high sensitivity, ALMA may however go beyond that and try to explore the centers of quasar host galaxies at high redshift. As the models presented here indicate, one may expect that the chemistry in these regions is dominated by X-ray emission from the central engine. In contrast to soft-UV photons, X-rays have high heating efficiencies and low efficiencies for dissociation. They can therefore give rise to an entirely different molecular cloud chemistry, and can excite high-J CO lines well above the (10-9) transition.

Additional probes for the central XDRs are available as well, in particular fine-structure lines like [OI] 63 μm, [OI] 146 μm and [CII] 158 μm. In particular for high X-ray luminosities, the fine-structure lines of [OI] 63 μm and [CII] 158 μm may provide a significant amount of flux even for gas densities of 10^4 cm^{-3}. In general, the fine-structure lines become optically thick more easily than the rotational CO lines, in particular at densities of 10^5 cm^{-3}. Therefore, they are almost insensitive to the column density, but quite sensitive to the X-ray flux. The CO lines, on the other hand, have a more non-trivial dependence on the X-ray flux and the column density, as in particular for high fluxes some dissociation effects will be present. If ideally both the CO lines and the fine-structure lines are observed, it may thus be possible to derive constraints both on column densities and the local X-ray flux. The line ratios may also provide valuable information on cloud gas density and temperature, as pointed out by Meijerink et al. [2007].

We have applied our models to available data for observed XDRs. For galaxies such as NGC 1068, we expect that the CO line intensity rises continuously up to the (17-16)

---

[1] http://www.jwst.nasa.gov/

7. PROBING HIGH-REDSHIFT QUASARS WITH ALMA[1]

transition. Such high excitation is not possible in PDR models [Spaans & Meijerink, 2008]. A detection of such lines therefore provides a unique diagnostic for an XDR.

We consider another example, the lensed galaxy APM 08279. There, the situation is not fully clear and is currently consistent with the presence of an XDR, which should also be expected due to the high X-ray luminosity, but also with the presence of a PDR with a radiation field of $G_0 = 10^{4.5}$. To some degree, this is due to uncertainties in the measured line fluxes. We expect that improved measurements with ALMA may help to distinguish these scenarios further and probe the local gas motions in more detail.

We finally discussed the chances to find additional sources with ALMA in a survey that scans a few arcmin^2. General considerations regarding the average accretion history indicate that quasar duty cycles should be larger at high redshift, so that black holes can grow to the required masses at $z = 6$. We therefore expect more sources compared to what one expects from a naive extrapolation of the observed quasar population. Our estimates indicate that about one source per arcmin^2 may be present near $z \sim 6$. At larger redshift, the number of sources may decrease significantly and one may need to rely on follow-up observations of sources detected by JWST. Due to the potentially high spatial resolution, the transport of gas to the center of the host galaxy may be probed for the first time in such galaxies even at $z \sim 10$.

We finally note that the diagnostics considered in this paper, in particular the high-J CO lines, may not only be important for high-redshift observations with ALMA, but also for future space-borne observations in the local universe with SPICA or FIRI.

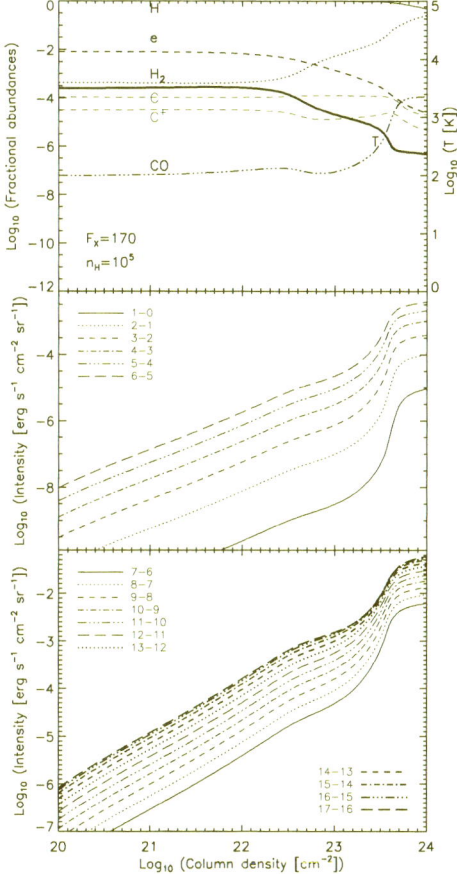

Figure 7.3: A model for the X-ray chemistry in NGC 1068. The adopted flux impinging on the cloud is 170 erg s^{-1} cm^{-2}. The adopted density is 10^5 cm^{-3}. Top: The abundances of different species as a function of column density. Middle: The low-J CO lines as a function of column density. Bottom: The high-J CO lines as a function of column density.

7. PROBING HIGH-REDSHIFT QUASARS WITH ALMA[1]

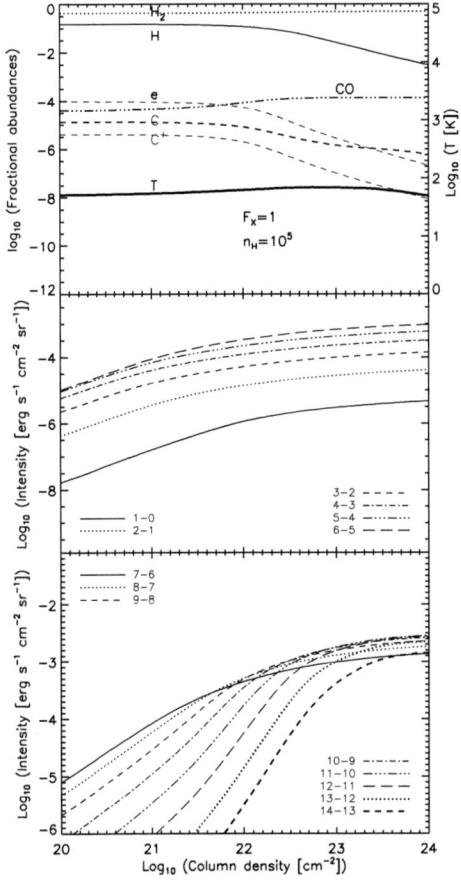

Figure 7.4: The X-ray chemistry in a system with X-ray flux of 1 erg s^{-1} cm^{-2} impinging on the cloud. The adopted density is 10^5 cm^{-3}. Top: The abundances of different species as a function of column density. Middle: The low-J CO lines as a function of column density. Bottom: The high-J CO lines as a function of column density.

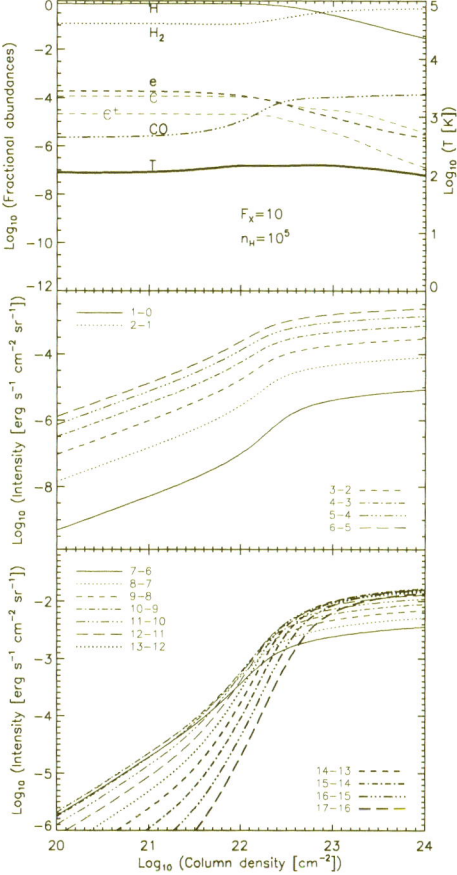

Figure 7.5: The X-ray chemistry in a system with X-ray flux of 10 erg s^{-1} cm^{-2} impinging on the cloud. The adopted density is 10^5 cm^{-3}. Top: The abundances of different species as a function of column density. Middle: The low-J CO lines as a function of column density. Bottom: The high-J CO lines as a function of column density.

7. PROBING HIGH-REDSHIFT QUASARS WITH ALMA[1]

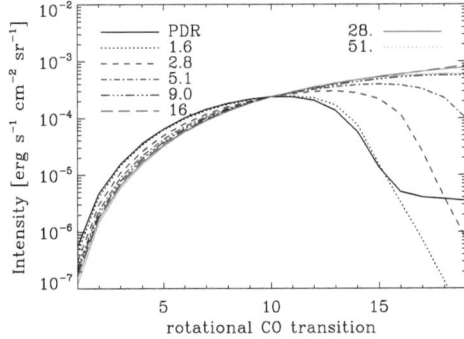

Figure 7.6: A comparison of the CO line SED in case of an intense starburst with $G_0 = 10^5$ with the corresponding SED for molecular clouds under X-ray irradiation, for different X-ray fluxes in erg s^{-1} cm^{-2}. The spectra are normalized such that they have the same intensity in the 10th transition. If the impinging X-ray flux is at least 2.8 erg s^{-1} cm^{-2}, observations of the 15th CO rotational transition can clearly discriminate between PDR and XDR chemistry.

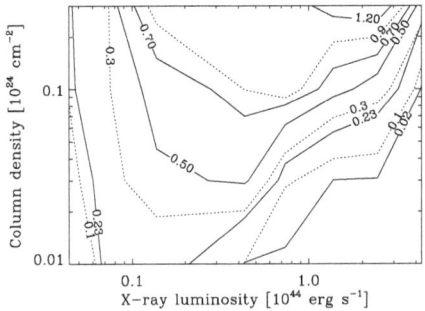

Figure 7.7: The expected flux in mJy for high-J CO lines, for a central XDR of 200 pc, and molecular clouds of 10^5 cm^{-3}, as a function of X-ray luminosity and cloud column density. We focus on lines that fall in ALMA band 6, which offers a good compromise between angular resolution and sensitivity. For a source at $z = 5$ (solid line), this corresponds to the (10-9) CO transition, for a source at $z = 8$, it corresponds to the (14-13) CO transition.

7.7 Conclusions

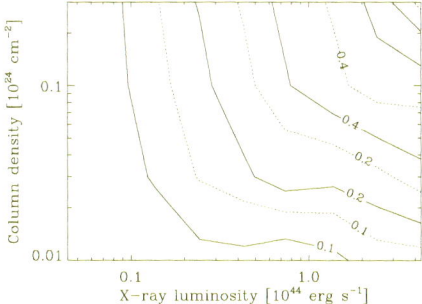

Figure 7.8: The expected flux in mJy for the [CII] 158 μm line, for a central XDR of 200 pc, and molecular clouds of 10^5 cm^{-3}, as a function of X-ray luminosity and cloud column density. The solid line corresponds to a source at $z = 5$, and the dashed line to a source at $z = 8$. At $z = 5$, the line is redshifted into ALMA band 7, with a sensitivity of 0.08 mJy (1 day, 3σ, 300 km/s). At $z = 8$, it falls into ALMA band 6 with a sensitivity of 0.04 mJy.

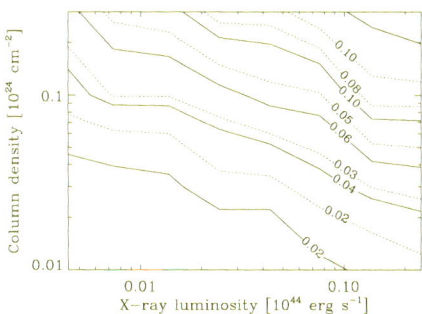

Figure 7.9: The expected flux in mJy for the [CII] 158 μm line, for a central XDR of 200 pc, and molecular clouds of 10^4 cm^{-3}, as a function of X-ray luminosity and cloud column density. The solid line corresponds to a source at $z = 5$, and the dashed line to a source at $z = 8$. At $z = 5$, the line is redshifted into ALMA band 7, with a sensitivity of 0.08 mJy (1 day, 3σ, 300 km/s). At $z = 8$, it falls into ALMA band 6 with a sensitivity of 0.04 mJy.

7. PROBING HIGH-REDSHIFT QUASARS WITH ALMA[1]

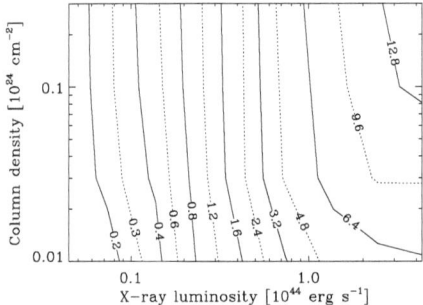

Figure 7.10: The expected flux in mJy for the [OI] 63 μm line, for a central XDR of 200 pc, and molecular clouds of 10^5 cm^{-3}, as a function of X-ray luminosity and cloud column density. The solid line corresponds to a source at $z = 5$, and the dashed line to a source at $z = 8$. At $z = 5$, the line is redshifted into ALMA band 10, with a sensitivity of 0.2 mJy (1 day, 3σ, 300 km/s). At $z = 8$, it falls into ALMA band 8 with a sensitivity of 0.13 mJy.

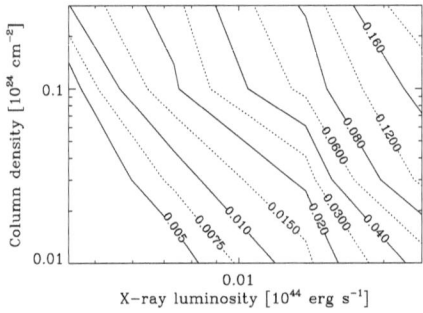

Figure 7.11: The expected flux in mJy for the [OI] 63 μm line, for a central XDR of 200 pc, and molecular clouds of 10^4 cm^{-3}, as a function of X-ray luminosity and cloud column density. The solid line corresponds to a source at $z = 5$, and the dashed line to a source at $z = 8$. At $z = 5$, the line is redshifted into ALMA band 10, with a sensitivity of 0.2 mJy (1 day, 3σ, 300 km/s). At $z = 8$, it falls into ALMA band 8 with a sensitivity of 0.13 mJy.

7.7 Conclusions

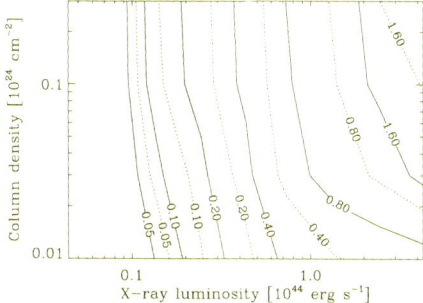

Figure 7.12: The expected flux in mJy for the [OI] 146 μm line, for a central XDR of 200 pc, and molecular clouds of 10^5 cm^{-3}, as a function of X-ray luminosity and cloud column density. The solid line corresponds to a source at $z = 5$, and the dashed line to a source at $z = 8$. At $z = 5$, the line is redshifted into ALMA band 7, with a sensitivity of 0.37 mJy (1 day, 3σ, 300 km/s). At $z = 8$, it falls into ALMA band 6 with a sensitivity of 0.225 mJy.

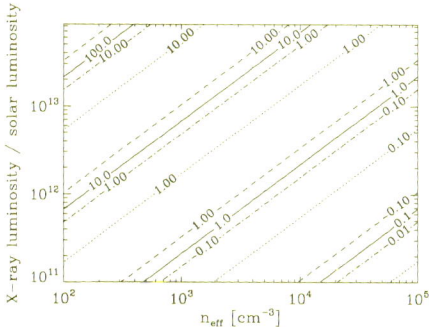

Figure 7.13: The expected flux in mJy for the CO (14-13) transition (solid line), [CII] 158 μm (dotted line), [OI] 63 μm (dashed line) and [OI] 146 μm (dot-dashed line) emission for a quasar at $z = 5$, as a function of X-ray luminosity and average density. We assume a typical cloud column density of 10^{23} cm^{-2}, and a soft-UV field $G_0 = 100$.

7. PROBING HIGH-REDSHIFT QUASARS WITH ALMA[1]

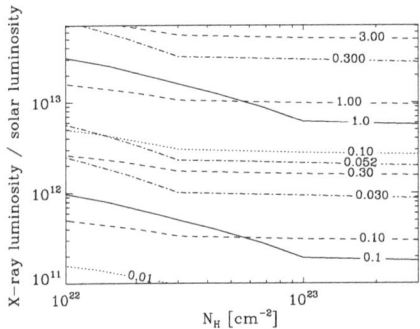

Figure 7.14: The expected flux in mJy for the CO (14-13) transition (solid line), [CII] 158 μm (dotted line), [OI] 63 μm (dashed line) and [OI] 146 μm (dot-dashed line) emission for a quasar at $z = 8$, as a function of X-ray luminosity and cloud column density. We assume a typical cloud density of 10^4 cm^{-3}, and a soft-UV field $G_0 = 100$.

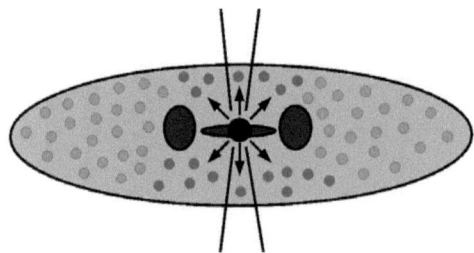

Figure 7.15: A sketch for a situation with an inhomogeneous XDR, motivated by the observations of Galliano et al. [2003] in NGC 1068. While X-rays are shielded by a central absorber along the line of sight, they may stimulate emission in molecular clouds in the perpendicular direction with the typical characteristics of XDRs.

7.7 Conclusions

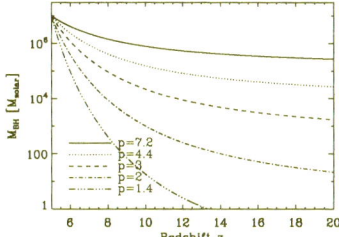

Figure 7.16: The average accretion history of a black hole with 10^7 $M_\odot$ at $z = 5$, depending on a coefficient p which is a function of the conversion of rest-mass energy to luminous energy, the Eddington ratio and the duty cycle.

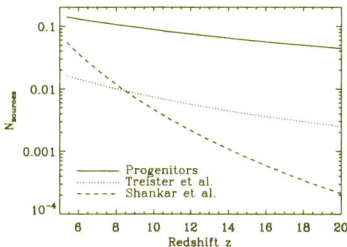

Figure 7.17: Estimates for the number of sources in a redshift interval of $\Delta z = 0.5$ within a solid angle of $(1')^2$. We show extrapolations of the results by Shankar et al. [2009] and Treister et al. [2009] to high redshift, for luminosities larger than 10^{44} erg s^{-1}. In addition, we give an estimate based on the number of high-redshift black holes required in order to produce the present-day 10^9 $M_\odot$ black holes. As discussed in the text, the actual number of black holes should lie in between these estimates.

7. PROBING HIGH-REDSHIFT QUASARS WITH ALMA[1]

8

Discussion and outlook

In the previous chapters, I have outlined observational constraints for the stellar population, primordial magnetic fields and dark matter models, as well as new opportunities that will help to explore the high-redshift universe in more detail. In this chapter, I will summarize the main results, outline the main open questions and provide an outlook for future work that will help to improve the previous results.

8.1 Summary

In this section, I summarize the main results regarding observational constraints and future perspectives that were found in this work.

8.1.1 Reionization constraints

Reionization constraints have been considered in Chapter 2 and 5 and provide upper limits on the existence and nature of additional physics as well as the stellar population. The main results are:

1. Pop. II stars alone cannot provide enough ionizing radiation to produce the required reionization optical depth.

2. Constraints on the magnetic field strength result in values between 0.7 – 3 nG, depending on the precise model for the stellar population.

3. For dark matter annihilations, the parameter combination $\langle\sigma v\rangle f_{abs} m_{DM}^{-1} \geq 3\times 10^{-33}$ cm^3 s^{-1} eV^{-1} can be ruled out at 3σ, where $\langle\sigma v\rangle$ is the thermally-averaged annihilation cross section, f_{abs} the fraction of annihilation energy absorbed into the IGM and m_{DM} the mass of the dark matter particle in eV.

8. DISCUSSION AND OUTLOOK

4. For dark matter decay, the parameter combination $t_X f_{\text{decay}}^{-1} < 3 \times 10^{23}$ s can be ruled out at 3σ, where t_X is the lifetime of the dark matter particle and f_{decay} the fraction of the decay energy absorbed into the IGM.

5. Dark stars with 1000 $M_\odot$, as suggested by Freese et al. [2008a], are difficult to reconcile with cosmic reionization, unless a rapid transition to a different star formation mode is assumed.

6. The same holds for dark star models which consider scattering between baryons and dark matter particles [Iocco et al., 2008; Freese et al., 2008c; Yoon et al., 2008]. These models tend to produce very long lifetimes of ~ 100 Myr or more, again overproducing the amount of UV photons.

Of course, it is important to assess the reliability of these conclusions. Points 3 and 4 are very reliable, as their derivation assumes that reionization is due to dark matter annihilation / decay alone. Point 2 has a weak dependence on the assumptions for the stellar population, but also appears to be well established. Point 5 appears rather firm, as well as point 6, though the latter might also be reinterpreted as a constraint for the scattering cross section between dark matter and baryons. Regarding point 1, one has to deal with two uncertainties: The star formation efficiency and the escape fraction of UV photons. In this work, the latter was chosen according to the observed escape fractions in high-redshift galaxies, while the star formation efficiency was chosen according to general considerations regarding the Jeans mass and the mass that goes into collapse [Abel et al., 2002; Bromm & Larson, 2004]. To alleviate this constraint, one would need to find a mechanism that increases the product of the two by two orders of magnitude to reproduce the required optical depth. However, there is no motivation for higher star formation efficiencies from simulations, and given that Pop. II stars produce considerably less UV photons per stellar baryon, it seems unlikely that they can photo-evaporate the halo in the same way as Pop. III stars.

8.1.2 21 cm observations

As shown in Chapter 3, primordial magnetic fields can be constrained further by future 21 cm observations, for instance with LOFAR or SKA. Such magnetic fields have mainly two effects on the 21 cm signal:

At redshifts $z > 40$, they increase the amplitude of the 21 cm power spectrum due to the additional heat input into the gas, which is due to ambipolar diffusion heating and decaying MHD turbulence. To measure this effect is however challenging, as the foreground emission corresponds to temperatures which are by orders of magnitude higher. On the other hand, as pointed out by Di Matteo et al. [2002]; Oh & Mack [2003]; Zaldarriaga et al. [2004]; Sethi

[2005]; Shchekinov & Vasiliev [2007], the foreground emission is expected to be featureless in frequency, which may allow a sufficiently accurate subtraction.

A more relevant effect is the dynamical impact of primordial fields on the formation of galaxies. As discussed in Chapters 2 and 3, the magnetic pressure can lead to very high Jeans masses and prevent gas collapse in the first minihalos. This provides an efficient mechanism to delay reionization and the formation of the first luminous sources. If 21 cm observations confirm that reionization started early, such scenarios would be ruled out as well, constraining the primordial magnetic field strength down to 0.1 nG.

Such observations also provide a relevant test for the effects of dark matter annihilation / decay in the early universe, as shown by Furlanetto et al. [2006b]; Chuzhoy [2008]; Cumberbatch et al. [2008]. More stringent upper limits are usually found from considerations of the cosmic gamma-ray background, as most of the annihilation products contribute to the cosmic backgrounds rather than being absorbed into the IGM [Cumberbatch et al., 2008]. The following subsection summarizes the main results from these considerations.

8.1.3 Constraints from the cosmic gamma-ray background and the atmospheric neutrino background

In Chapter 5 and 6, I used the cosmic gamma-ray background and the atmospheric neutrino background to derive further constraints on dark matter models. In a similar way, as shown in Chapter 7, one can also use the unresolved X-ray background to constrain the black hole population at high redshift. In summary, I found the following results:

1. Constraints from the unresolved X-ray background suggest that one could find up to ~ 10 black holes of 10^6 $M_\odot$ in a field of view of $(1')^2$ between redshift 6 and 10, considering a typical redshift interval of $\Delta z = 0.5$.

2. Light dark matter models with constant thermally-averaged annihilation cross sections (s-wave) are ruled out due to the observed gamma-ray background.

3. For heavy dark matter candidates with annihilation cross sections at the natural scale and branching-ratio of 1/3 for gamma-rays, as typically considered in dark star models, particle masses smaller than 30 GeV are ruled out.

4. If neutrinos are generated as annihilation products, their contribution is well below the atmospheric neutrino flux, thus providing no significant constraint on the models.

5. In the centers of primordial halos with virial temperatures smaller than 10^4 K, the steep cusp in the dark matter density profile can potentially contribute to the cosmic gamma-ray background, but this effect is very uncertain.

8. DISCUSSION AND OUTLOOK

Again, these results are rather robust. The first conclusion depends on assumptions regarding the black hole spectra, but holds for a reasonable range of parameters. It is particularly promising regarding the search for high-redshift quasars with ALMA and JWST, as it provides an upper limit that is larger than other estimates for the black hole population, thus admitting a potentially large number of sources. Other estimates taking into account the average accretion history of black holes predict smaller numbers, that are however still sufficient to find sources in a solid angle comparable to the Hubble Deep Field.

The second result is of course sensitive to assumptions regarding the clumpiness of dark matter, but they were chosen to be very conservative and can be relaxed even further. While similar results had been obtained previously from Galactic observations [Boehm et al., 2004b], our results are less sensitive to model assumptions regarding the details of the Milky Way, and only make use of the observed spectral shape of the emitted radiation. Therefore, s-wave annihilation for light dark matter is now firmly ruled out. The third conclusion slightly depends on the assumption regarding the clumpiness of dark matter at early times, but was also chosen to be conservative. It holds only for dark matter models considered in dark star scenarios, where it is assumed that a fraction of $\sim 1/3$ of the annihilation energy goes into gamma-rays. It therefore provides a lower limit for the dark matter particle mass in such scenarios. In more generic cases with smaller branching ratios to gamma-rays, this constraint is relaxed considerably [Mack et al., 2008]. This is quite similar for conclusion 5, while conclusion 4 is hardly sensitive to model assumptions.

8.1.4 Signatures of primordial molecules in the CMB

As an additional way to explore the early universe, I have considered the possibility to search for signatures of primordial molecules in the CMB (see Chapter 4). This was in particular motivated by a previous calculation of Black [2006] which indicated that the signatures of H^- could be found with modern technology. I therefore performed a self-consistent calculation of the chemistry in the early universe to investigate the abundances of the first molecules, and calculated the signatures for the most important species. The results showed that Black [2006] largely overestimated the signature from H^-. If the inverse reactions are taken into account, the net effect drops even further. It is thus considerably below the limit where it becomes observable. Also other species like HeH^+ and HD^+ would need to increase by orders of magnitude in their abundance to be detectable with the Planck satellite.

8.1.5 High redshift quasar observations with ALMA

I examined the question how future observations may help to understand the origin of the first supermassive black holes. At $z \sim 6$, black holes with 10^9 $M_\odot$ have been discovered, and their origin is still controversial (see Chapter 7). Even at those redshifts, their host galaxies

8.1 Summary

were found to have supersolar metallicities [Pentericci et al., 2002], and also the most recent data show no evolution of quasar metallicity with redshift [Juarez et al., 2009]. Here, I summarize the main results regarding the detectability of their progenitors at higher redshift, and in particular the fluxes from the central X-ray dominated region. A measurement of those would allow to probe gas close to the black hole, and thus to better constrain models for the circumnuclear disk.

This work has been inspired during a visit of Marco Spaans in Heidelberg, who presented first estimates on the expected fluxes in high-redshift quasars with low-metallicity gas [Spaans & Meijerink, 2008]. Our study was based on a semi-analytic model for the size of the central X-ray dominated region and chemical models provided by Meijerink & Spaans [2005]. We have also compared these models to known systems, in particular NGC 1068, APM 08279+5255 and SDSS J114816.64+525150.3. We found the following results:

1. For metallicities of at least 10^{-1} $Z_\odot$, the gas cooling is clearly dominated by metals, and most of the flux is expected in CO, O and C^+. Sources at $z \sim 8$ can be detected with ALMA within less than an hour, for black hole masses of $\sim 10^7$ $M_\odot$.

2. As the main coolants can be observed, this provides a good estimate for the total heating (=cooling) rate in the dense clouds and the corresponding X-ray luminosity.

3. The CO luminosity depends both on X-ray luminosity and column density, while the flux in [CII] and [OI] is found to be optically thick in dense gas and depends mostly on the X-ray luminosity.

4. In the central X-ray dominated regions, high temperatures of ~ 1000 K may be present in molecular clouds, leading to a flat CO SED at high rotational transitions. Measurements of the $J > 10$ lines are particularly valuable to discriminate from PDR models.

5. Signatures of X-ray dominated chemistry have been found both in NGC 1068, APM 08279+5255 and the Cloverleaf quasar. These are also consistent with the X-ray flux from the central black hole. Extreme PDR cases are however not yet ruled out for the latter two cases, as no $J > 10$ CO lines have been observed yet. With ALMA, this will be possible, and we expect that the two cases can be clearly discriminated.

6. So far, SDSS J114816.64+525150.3 has only been probed on large scales, where we find typical signatures of PDRs. Due to the presence of [CI], [CII] and CO lines, we can probe both the low- and high-density gas. Interestingly, the high-density gas seems to be exposed to a stronger radiation field. This may indicate that star formation in this system started just recently.

8. DISCUSSION AND OUTLOOK

7. The expected number of sources is quite uncertain and model-dependent. In particular, evolutionary changes in parameters like the duty cycle and the Eddington efficiency need to be better understood. In an area of $(1')^2$, we expect that up to one detectable source will be present.

ALMA line observations provide valuable diagnostics to derive the gas temperature and density [Meijerink & Spaans, 2005]. A detection of the various different ALMA observables may yield further information regarding the gas composition and metal abundance ratios. If the broad-line region is not obscured, the velocity dispersion of the corresponding line fluxes yields an upper limit to the black hole mass. This can be compared with the X-ray luminosity inferred from ALMA observations in order to constrain the Eddington ratio at high redshift. The velocity dispersion of the ALMA line fluxes provides detailed information about the gas dynamics and the dynamical mass. With an upper limit from the black hole mass, this can be used to constrain the Magorrian relation at high redshift.

8.2 Open questions

In this section, I discuss some of the main open questions regarding the early universe, both concerning the physics, the reionization process, primordial star formation and the formation and evolution of high-redshift quasars. These may provide a starting point for future investigations.

8.2.1 Physics of the early universe

At present, even if the most optimistic constraints from Chapter 2 on primordial magnetic fields hold, they hardly constrain any of the scenarios discussed in Chapter 1.2.1.2, perhaps with the exception of some inflationary scenarios, which can predict a huge range of different values. However, a comparison with the calculation of Banerjee & Jedamzik [2003] shows that the current constraints are roughly comparable to the field strengths that could be expected from the QCD phase transition, in particular for those models which assume high helicity. High field strengths are also predicted in case of hydrodynamic instabilities during the QCD phase transitions [Sigl et al., 1997]. Stronger observational constraints on the magnetic field strength will therefore provide direct constraints for the physics at the QCD phase transition.

From a theoretical point of view, it is necessary to study the impact of such magnetic fields on the first generation of stars in more detail to better understand their role in the early universe and to derive potential constraints. In particular, they may not only change the chemistry, but also affect the dynamics of the gas during collapse. An outlook of the currently ongoing work regarding this topic is given below.

The question regarding the true nature of dark matter is also unresolved, as well as its potential annihilation and decay channels. Currently, the FERMI satellite [1] searches for signatures of dark matter annihilation / decay in the gamma-ray background and the centers of galaxies. A positive detection would certainly stimulate the modeling of dark matter and make it possible to specifically explore the effects of the corresponding scenarios in the early universe. In this case, the effects of dark matter annihilation on the first stars would need to be developed with much more detail, to take into account the detailed dynamics of the collapse process. The currently proposed theories vary strongly in the expected range of stellar masses and also concerning the lifetime of the "dark" phase. It may thus be important to address these questions with numerical simulations that account for the dynamics, the chemistry as well as the dark matter heating effects.

8.2.2 Reionization and the stellar population

Even in the absence of the primordial magnetic fields and dark matter annihilation / decay, many questions regarding reionization and stellar populations are still open. This concerns in particular the transition to the star formation mode in the present-day universe. A number of different feedback mechanisms that may contribute to this transition have already been discussed in the introduction. The main ingredient is generally assumed to be metallicity [Bromm et al., 2001; Schneider et al., 2003; Clark et al., 2008], though it was also suggested that magnetic fields may be responsible for this transition [Silk & Langer, 2006]. In order to understand the transition to a different star formation mode, one needs to understand how metal enrichment proceeds in the early universe. Simulations by Tornatore et al. [2007] show that metal enrichment is a highly inhomogeneous process, and pristine bubbles of primordial gas may even be left near $z \sim 2$. This is encouraging regarding the search for Pop. III stars. Metal enrichment in the early universe will be explored in more detail with a new algorithm for chemical mixing in hydrodynamic simulations [Greif et al., 2009].

This is particularly relevant for star formation in atomic cooling halos. Even in the primordial case, they might harbor a different star formation mode than the first minihalos, as they consist of a rather turbulent two-phase medium [Greif et al., 2008]. The gas in the hot phase has a larger ionization degree and therefore forms coolants like H_2 and HD more efficiently, such that the gas can ultimately reach lower temperatures when it collapses. Accordingly, these will be reflected in smaller Jeans masses. The fragmentation will be triggered further due to the presence of turbulence, as found in simulations of Clark et al. [2008]. Of course, in the presence of metals these effects may be enhanced even further [Schneider et al., 2003; Clark et al., 2008; Omukai et al., 2008]. This question therefore needs to be addressed with self-consistent calculations of star formation and metal-enrichment in the first galaxies.

---

[1] http://www.nasa.gov/mission_pages/GLAST/main/index.html

8. DISCUSSION AND OUTLOOK

As mentioned in the introduction, the effects of magnetic fields generated during protostellar collapse are currently not well understood. Some simulations are available by Machida et al. [2006], but with initial conditions corresponding to present-day star formation rather than the cosmological context in the early universe. The results are thus very similar to corresponding works in present-day star formation [Ziegler & Rüdiger, 2000; Price & Bate, 2007; Hennebelle & Fromang, 2008]. On the other hand, Xu et al. [2008] calculated the generation of magnetic fields due to the Biermann battery mechanism [Biermann, 1950], but could not resolve the disks in which the field strength may be amplified by dynamo effects. To perform simulations that bridge these scales is thus one major challenge for the future.

In addition, it is also clear that magnetic fields will be generated in the first stars or even during their formation (see outlook). Magnetic winds or supernova explosions will bring them in the environment and ultimately turn it into a magnetized medium [Rees, 1987]. Small-scale turbulence from supernova explosions may contribute to the formation of magnetic fields that can affect the following stellar generations [Kandus et al., 2004]. Such fields can affect the chemistry during collapse by ambipolar diffusion, but also the dynamics. This form of magnetic feedback still needs to be explored in the standard framework of reionization.

8.2.3 High-redshift quasars

The main open question regarding the evolution of high-redshift quasars have already been discussed in detail in the introduction and in Chapter 7. I will therefore just briefly summarize the main questions that need to be resolved:

- What is the origin of 10^9 $M_\odot$ black holes at $z \sim 6$?

- How do such quasars build up supersolar metallicities?

- Should we expect different metallicities at higher redshift?

- How does the Magorrian relation evolve with redshift?

As shown in Chapter 7, some of these questions can be addressed observationally in the era of ALMA and JWST. In addition, improvements in theoretical models may help to confirm or rule out some scenarios based on consistency arguments, or provide predictions that can be used to distinguish between different scenarios.

8.3 Outlook

In this section, I will sketch current and future projects that will help to understand the early universe in more detail, with the goal to arrive at more detailed predictions that can be compared with current and future observations. Particularly promising is the launch of Planck in April this year (2009), which will provide better constraints about cosmic reionization on a short timescale. On longer timescales, a first set of ALMA antennas will be completed in about three years, allowing a first test of the predictions in Chapter 7 for instance by observing known quasars at high redshifts. In five years, ALMA should be complete and may, ideally in combination with JWST, start probing the first quasars at high redshift.

8.3.1 Magnetic fields

To understand the effects of primordial magnetic fields in more detail, I have calculated the effect of ambipolar diffusion heating on the gas chemistry during the collapse of primordial gas. This is based on a one-zone model by Glover & Savin [2009] which includes all relevant chemical processes down to number densities of 10^{14} cm^{-3}. These calculations indicate that comoving field strengths of 0.1 nG can change the evolution of the gas temperature in this density range, and weaker fields may become relevant at later stages. During the collapse, we find that the gas initially cools more rapidly, but starting from a higher temperature, as the IGM was heated by ambipolar diffusion. They reach roughly the same temperature minimum as in the non-magnetized case, so that the fragmentation mass scale would be unchanged. Subsequently, the gas temperature is however higher because of ambipolar diffusion. Thus, the protostellar accretion rate may be larger by a factor of a few. This may exceed the mass loss due to jets and outflows predicted by Machida et al. [2006].

In addition, we need to assess the possible role of dynamos in case primordial magnetic fields are weak. Dynamo theory was recently reviewed by Brandenburg & Subramanian [2005]. Particular efficient in the presence of turbulence might be the small-scale dynamo originally proposed by Kazantsev [1968], which leads to an exponential growth of the magnetic field during the eddy-turnover time. On small scales, the latter is significantly smaller than the free-fall time, allowing for rapid growth of the magnetic field until it saturates. The resulting magnetic field structure is expected to be highly tangled and intermittent. Such magnetic fields may help to transport angular momentum more efficiently and thus suppress the amount of fragmentation in the gas. After the formation of a disk, the $\alpha\omega$ dynamo may start operating and generate a magnetic field that is more coherent, and potentially able to drive jets and outflows. We plan to study such scenarios in more detail both with an analytic and a numerical approach. In this context, analytic approaches have the advantage that they can bridge a large range of scales and in particular treat those scales where the eddy turnover time is much smaller than the free-fall time. Numerical studies, on the other hand, may give a

8. DISCUSSION AND OUTLOOK

more realistic impression on the amount of turbulence present on the scales that are actually resolved, and can also account for other dynamical effects like the formation of a disk.

In pure hydrodynamical simulations, the presence of sub-sonic turbulence has already been confirmed by Abel et al. [2002] for high-redshift minihalos that form the first stars, finding that turbulence forms due to sub-sonic shocks during the collapse and regulating the transport of angular momentum. In the first galaxies, this turbulence was found to be supersonic with Mach numbers of order 10 [Wise & Abel, 2007a; Greif et al., 2008]. It therefore seems likely that the requirements for the small-scale dynamo are fulfilled in the first objects. If this can be confirmed, the consequences of such magnetic fields need to be studied further, in particular concerning the fragmentation properties of the protostellar disk.

8.3.2 Dark matter annihilation

The effects of dark matter annihilation on the formation of dark stars have already been explored in some detail, but important considerations are currently missing. For instance, most of the previous works focused on the later collapse phase in which dark matter annihilation heating may match the total cooling rate and therefore stop further collapse. However, very interesting effects can occur well before that phase. Prelimary results indicate that even a modest increase of the ionization fraction in the IGM due to annihilation of dark matter would considerably enhance the ability of the gas to form H_2 and HD and thus cool down to temperatures close to the CMB. This has a strong effect on the thermal Jeans mass and may even trigger fragmentation well before the time a dark star is expected to form. Similar results have also been found by Ripamonti et al. [2010].

It also needs to be assessed in more detail whether dark matter heating can indeed stop the collapse, or if it only leads to an increase of the gas temperature until the cooling rate is again larger than the heating rate. It seems nevertheless likely that heating due to dark matter annihilation would influence the gas during the collapse phase and may thus change the fragmentation behaviour and the initial mass function of the first stars. This can be addressed in more detail with numerical simulations that include hydrodynamics, primordial chemistry and dark matter heating. In addition, it needs to be explored under which conditions dark matter cusps can form that give rise to a sufficient amount of heating, i.e. whether this is still possible in the presence of turbulence and supersonic inflows that have been found in the first galaxies [Wise & Abel, 2007a; Greif et al., 2008], and how the presence of a non-zero metallicity would influence this scenario.

8.3.3 The first quasars

In this work, I have provided estimates for the observables of high-redshift quasars with respect to ALMA based on models for the chemistry in X-ray dominated regions provided

8.3 Outlook

by Meijerink & Spaans [2005]; Meijerink et al. [2007]. These calculations have established that one may expect to observe black holes of $\sim 10^7$ $M_\odot$ with these telescopes even at $z \sim 10$ and beyond. This would be particularly important if surveys performed with JWST and other telescopes can provided a catalogue of potential sources. The high resolution of ALMA will in particular allow to probe the central circumnuclear disks around the supermassive black hole, in which the chemistry may be driven by the absorption of X-rays. as discussed in Chapter 7. This prediction may be directly tested by measuring the CO line SED in the central region. Models for the dynamics in these central regions have been developed by Thompson et al. [2005] and Kawakatu & Wada [2009]. In the picture of Kawakatu & Wada [2009], star formation and supernova feedback is required to drive turbulence and enhance the accretion, while Thompson et al. [2005] argue that in case of star formation, less mass is available for accretion. In practise, it seems likely that the competition between the star formation rate and the accretion rate is important and that they influence each other.

In both models, several properties of these circumnuclear disk appear similar to local starburst galaxies like Arp 220, in particular the average density, the spatial extent and the star formation rate [Greve et al., 2009]. Such intense star forming galaxies may therefore be an important probe for the conditions around a rapidly accreting black hole. The analogy is however not complete, as in the latter case, strong emission in the X-rays is expected that should influence the chemistry in molecular clouds. Based on Zeeman measurements, it has been shown that strong magnetic fields are present in the star-forming disk of Arp 220, of the order of a few mG. The implications of such magnetic fields for star formation and accretion have not yet been considered in theoretical models, and their origin is not yet explained. One might therefore speculate that similarly strong magnetic fields are present in the circumnuclear disks around supermassive black holes. In this case, they could significantly influence the dynamical evolution. In future studies, the relative importance of gravitational instabilities, supernova feedback, chemical feedback (X-rays vs starburst) and magnetic fields therefore needs to be addressed. The formation and growth of black holes will therefore remain an exciting field of research in the coming years.

8. DISCUSSION AND OUTLOOK

Appendix A

Free-free transitions involving H^- [1]

This appendix[1] provides additional material to Chapter 4. While the importance of free-free transitions is well-known in stellar atmospheres, this process is usually not considered in colder environments. Thus, high precision calculations are only available for stellar temperatures between 1000 and 10000 K [John, 1988; Bell & Berrington, 1987; Gingerich, 1961]. The free-free absorption coefficient k_{ff} is normalized per hydrogen atom and per unit electron pressure, and contributes to the optical depth by $d\tau = k_{ff} n_H n_e kT ds$. As the fits by John [1988] and Gingerich [1961] diverge at low temperatures, thus giving an unphysical high contribution at low redshifts, we calculate the absorption coefficient based on the formalism of Dalgarno & Lane [1966], which takes into account only the contribution of the leading term to the transition moment. This approximation is valid at low energies and is thus a reasonable choice for the low-temperature regime. Assuming that the initial electron energies are described by a Maxwell distribution, the free-free coefficient k_{ff} is given as

$$
\begin{aligned}
k_{ff}(\nu) &= 9.291 \times 10^{-3} T^{-5/2} (1 - e^{-a}) \int_0^\infty E^{-1} \left(\frac{E}{h_p \nu}\right)^3 \left(1 + \frac{h_p \nu}{E}\right)^{1/2} \\
&\quad \times\; e^{-aE/h_p \nu} \left(\left(1 + \frac{h_p \nu}{E}\right) q_0(E) \right. \\
&\quad +\; \left. q_0(E + h_p \nu)\right) dE \; \text{cm}^4 \; \text{dyne}^{-1},
\end{aligned} \qquad (A.1)
$$

where ν is the photon frequency, $q_0(E)$ the zero-order elastic scattering cross section and $a = h_p \nu / kT$ a dimensionless parameter. While Dalgarno & Lane [1966] used effective range theory [O'Malley et al., 1961] to expand the cross section, we use the more accurate result

---

[1]This appendix is based on the appendix of Schleicher et al., A&A, 490, 521, 2008, reproduced with permission © ESO.

A. FREE-FREE TRANSITIONS INVOLVING H⁻ [1]

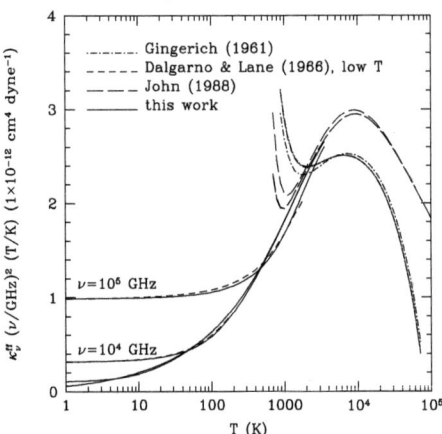

Figure A.1: The free-free absorption coefficient of H⁻ for 10^2, 10^3, 10^4 and 10^5 GHz, as a function of temperature. Given are the fits of John [1988] and Gingerich [1961] for the high-temperature regime, the calculation of Dalgarno & Lane [1966] based on effective range theory as well as the new calculation of this work for the low-temperature regime up to 2000 K.

of Dalgarno et al. [1999] as fitted by Pinto & Galli [2008], given by

$$q_0(E) = \frac{4 \times 10^{-15} \text{ cm}^2}{(1 + E/(3.8\text{eV}))^{1.84}}. \tag{A.2}$$

In the range between 1 and 10^7 GHz and between 0.1 and 2000 K, our results are well-fit by the expression

$$k_{ff}(\nu) = 10^{(a_1+a_2x+a_3x^2)(b_1+b_2y+b_3y^2)} \text{ cm}^4 \text{ dyne}^{-1}, \tag{A.3}$$

where $x = \log_{10}(\nu/\text{GHz})$ and $y = \log_{10}(T/\text{K})$, and $a_1 = -3.2421$, $a_2 = -0.502052$, $a_3 = 0.0117164$, $b_1 = 4.05293$, $b_2 = 0.169299$, $b_3 = -0.00548517$. At stellar temperatures, our results differ by about 10% from the calculation of Bell & Berrington [1987], which is based on a more detailed treatment, whereas at lower temperatures, we expect an even higher accuracy of our result. We thus adopt the fit of John [1988] to the calculation of Bell & Berrington [1987] for $T > 2000$ K and Eq. (A.3) for lower temperatures. The different calculations are compared in Fig. A.1.

A. FREE-FREE TRANSITIONS INVOLVING H[-] [1]

Appendix B

Reaction rates for primordial chemistry[1]

The table[1] on the next page summarizes the chemical model developed in Chapter 4. The main differences compared to previous works like Galli & Palla [1998]; Stancil et al. [1998] have been discussed in Chapter 4.4.

---

[1]This appendix is based on Schleicher et al., A&A, 490, 521, 2008, reproduced with permission © ESO.

Table B.1: Collisional and radiative rates.

Reaction number	Reaction	Rate [cgs]	Reference
1	$H + e \to H^- + \gamma$	$1.4 \times 10^{-18} T^{0.928} \exp(-T/16200)$	GP98
2	$H^- + H \to H_2 + e$	1.5×10^{-9} for $T < 300$	GP98
		$4.0 \times 10^{-9} T^{-0.17}$ for $T > 300$	GP98
3	$H + H^+ \to H_2^+ + \gamma$	$\text{dex}(-19.38 - 1.523 \log_{10}(T) + 1.118 \log_{10}^2(T)$	GP98
		$-0.1269 \log_{10}^3(T))$	
4	$H_2^+ + H \to H_2 + H^+$	6.4×10^{-10}	GP98
5	$2H + H \to H_2 + H$	$5 \times 10^{-29} T^{-1}$ for $T < 300$	PSS83
6	$H_2 + H^+ \to H_2^+ + H$	see reference	SKHS04b
7	$H_2 + e \to 2H + e$	$1.91 \times 10^{-9} T^{0.136} \exp(-53407.1/T)$	TT02
8	$H_2 + H \to 3H$	Fit to data of MSM96	MSM96
9	$H^- + e \to H + 2e$	see reference	AAZN97
10	$H^- + H \to 2H + e$	see reference	AAZN97
11	$H^- + H^+ \to 2H$	$1.40 \times 10^{-7} (T/300)^{-0.487} \exp(T/29300)$	LSD02
12	$H^- + H^+ \to H_2^+ + e$	$6.9 \times 10^{-9} T^{-0.35}$ for $T < 8000$	GP98
		$9.6 \times 10^{-7} T^{-0.9}$ for $T > 8000$	GP98
13	$H_2^+ + e \to 2H$	$2.0 \times 10^{-7} T^{-0.5}$	GP98
14	$H_2^+ + H^- \to H + H_2$	$5 \times 10^{-6} T^{-0.5}$ for $T > 100$	AAZN97
15	$H_2 + e \to H + H^-$	$3.67 \times 10^1 T^{-2.28} \exp(-\frac{47172}{T})$	CCDL07
16	$H^- + \gamma \to H + e$	$1.1 \times 10^{-1} T_r^{2.13} \exp(-8823/T_r)$	GP98
17	$H_2^+ + \gamma \to H + H^+$	$1.63 \times 10^7 \exp(-32400/T_r)$	GP98
18	$H_2 + \gamma \to H_2^+ + e$	$2.9 \times 10^2 T_r^{1.56} \exp(-178500/T_r)$	GP98

Table B.1: Collisional and radiative rates in cgs units.

Reaction number	Reaction	Rate [cgs]	Reference
19	$H_2^+ + \gamma \rightarrow 2H^+ + e$	$9.0 \times 10^1 T_r^{1.48} \exp(-335000/T_r)$	GP98
20	$H_2 + \gamma \rightarrow (H_2)^* \rightarrow 2H$	$1.13 \times 10^6 T_r^{0.369} \exp(-140000/T_r)$	GJ07
21	$D + \gamma \rightarrow D + e$	estimated by rate 16	
22	$HD^+ + \gamma \rightarrow D + H^+$	estimated by half of rate 17	
23	$HD^+ + \gamma \rightarrow H + D^+$	estimated by half of rate 17	
24	$HD + \gamma \rightarrow H^+ + D^+ + e$	estimated by rate 19	
25	$HD + \gamma \rightarrow HD^+ + e$	estimated by rate 18	
26	$D + e \rightarrow D + \gamma$	$3.6 \times 10^{-12} (T/300)^{-0.75}$	SLD98
27	$D + H^+ \rightarrow D^+ + H$	$2 \times 10^{-10} T^{0.402} \exp(-37.1/T) - 3.31 \times 10^{-17} T^{1.48}$	Savin (2002)
28	$D^+ + H \rightarrow D + H^+$	$2.06 \times 10^{-10} T^{0.396} \exp(-33.0/T) + 2.03 \times 10^{-9} T^{-0.332}$	Savin (2002)
29	$D + H \rightarrow HD + \gamma$	see reference	Dickinson (2005, 2008)
30	$D + H_2 \rightarrow HD + H$	$1.69 \times 10^{-10} \exp(-4680/T + 198800/T^2)$ for $T > 200$	GP02
		$9.0 \times 10^{-11} \exp(-3876/T)$ for $T < 200$	GP98
31	$HD^+ + H \rightarrow HD + H^+$	6.4×10^{-10}	SLD98
32	$D^+ + H_2 \rightarrow HD + H^+$	$1.0 \times 10^{-9}(0.417 + 0.846 \log_{10}(T) - 0.137 \log_{10}^2(T))$	GP02
33	$HD + H \rightarrow D + H_2$	$5.25 \times 10^{-11} \exp(-4430/T + 173900/T^2)$ for $T > 200$	GP02
		$3.2 \times 10^{-11} \exp(-3624/T)$ for $T < 200$	GP98
34	$HD + H^+ \rightarrow D^+ + H_2$	$1.1 \times 10^{-9} \exp(-488/T)$	GP02
35	$D + H^+ \rightarrow HD^+ + \gamma$	$\text{dex}(-19.38 - 1.523 \log_{10}(T) + 1.118 \log_{10}^2(T) - 0.1269 \log_{10}^3(T))$	GP98
36	$D^+ + H \rightarrow HD^+ + \gamma$	$\text{dex}(-19.38 - 1.523 \log_{10}(T) + 1.118 \log_{10}^2(T) - 0.1269 \log_{10}^3(T))$	GP98

Table B.1: Collisional and radiative rates in cgs units.

Reaction number	Reaction	Rate [cgs]	Reference
37	$HD^+ + e \rightarrow D + H$	$7.2 \times 10^{-8} T^{-0.5}$	SLD98
38	$D + e \rightarrow D^- + \gamma$	$3.0 \times 10^{-16}(T/300)^{0.95} \exp(-T/9320)$	SLD98
39	$D^+ + D^- \rightarrow 2D$	$1.96 \times 10^{-7}(T/300)^{-0.487} \exp(T/29300)$	LSD02
40	$H^+ + D^- \rightarrow D + H$	$1.61 \times 10^{-7}(T/300)^{-0.487} \exp(T/29300)$	LSD02
41	$H^- + D \rightarrow H + D^-$	$6.4 \times 10^{-9}(T/300)^{0.41}$	SLD98
42	$D^- + H \rightarrow D + H^-$	$6.4 \times 10^{-9}(T/300)^{0.41}$	SLD98
43	$D^- + H \rightarrow HD + e$	$1.5 \times 10^{-9}(T/300)^{-0.1}$	SLD98
44	$D + H^- \rightarrow HD + e$	estimated by rate 42	This work
45	$H^- + D^+ \rightarrow D + H$	$1.61 \times 10^{-7}(T/300)^{-0.487} \exp(T/29300)$	LSD02
46	$He^{++} + e \rightarrow He^+ + \gamma$	$3.36 \times 10^{-10} T^{-0.5}(T/1000)^{-0.2}\left(1 + (T/10^6)^{0.7}\right)^{-1}$	Cen (1992)
47	$He^+ + \gamma \rightarrow He^{++} + e$	see reference	AAZN97
48	$He^+ + e \rightarrow He + \gamma$	see reference	AAZN97
49	$He + \gamma \rightarrow He^+ + e$	see reference	AAZN97
50	$He + H^+ \rightarrow He^+ + H$	$4.0 \times 10^{-37} T^{4.74}$ for $T > 10000$	GP98
		$1.26 \times 10^{-9} T^{-0.75} \exp(-127500/T)$ for $T < 10000$	GJ07
51	$He^+ + H \rightarrow He + H^+$	$1.25 \times 10^{-15} \times (T/300)^{0.25}$	ZDKL89
52	$He + H^+ \rightarrow HeH^+ + \gamma$, rad. ass.	$8.0 \times 10^{-20}(T/300)^{-0.24} \exp(-T/4000)$	SLD98
53	$He + H^+ + \gamma \rightarrow HeH^+ + \gamma$, stim. rad. ass.	$3.2 \times 10^{-20} T^{1.8}/(1 + 0.1 T^{2.04})$ $\times \exp(-T/4000)(1 + 2 \times 10^{-4} T_r^{1.1})$	JSK95, ZSD98
54	$He + H_2^+ \rightarrow HeH^+ + H$	$3.0 \times 10^{-10} \exp(-6717/T)$	GP98
55	$He^+ + H \rightarrow HeH^+ + \gamma$	$4.16 \times 10^{-16} T^{-0.37} \exp(-T/87600)$	SLD98
56	$HeH^+ + H \rightarrow He + H_2^+$	$0.69 \times 10^{-9}(T/300)^{0.13} \exp(-T/33100)$	LJB95

Table B.1: Collisional and radiative rates in cgs units.

Reaction number	Reaction	Rate [cgs]	Reference
57	$HeH^+ + e \rightarrow He + H$	$3.0 \times 10^{-8} (T/300)^{-0.47}$	SLD98
58	$HeH^+ + \gamma \rightarrow He + H^+$	$220 T_r^{0.9} \exp(-22740/T_r)$	JSK95
59	$HeH^+ + \gamma \rightarrow He^+ + H$	$7.8 \times 10^3 T_r^{1.2} \exp(-240000/T_r)$	GP98

In the table above, T denotes the gas temperature in Kelvin, T_e, the gas temperature in eV, T_r the temperature of radiation in K. dex(x) = 10^x. Acronyms: AAZN97: Abel, Anninos, Zhang, & Norman [1997], CCDL07: Capitelli, Coppola, Diomede, & Longo [2007], GJ07: Glover & Jappsen [2007], GP98: Galli & Palla [1998], GP02: Galli & Palla [2002], JSK95: Jurek, Špirko, & Kraemer [1995], LJB95: Linder, Janev, & Botero [1995], LSD02: Lepp, Stancil, & Dalgarno [2002], MSM96: Martin, Schwarz, & Mandy [1996], PSS83: Pequignot, Petitjean, & Boisson [1991], SKHS04b: Savin, Krstić, Haiman, & Stancil [2004], SLD98: Stancil, Lepp, & Dalgarno [1998], TT02: Trevisan & Tennyson [2002], ZDKL89: Zygelman, Dalgarno, Kimura, & Lane [1989], ZSD98: Zygelman, Stancil, & Dalgarno [1998]

Acknowledgement

For the excellent support during the work on this thesis, I would like to thank my supervisors Prof. Dr. Ralf Klessen and Prof. Dr. Matthias Bartelmann. I am further in debth to Dr. Robi Banerjee, Prof. Dr. Max Camenzind, Prof. Dr. Daniele Galli, Dr. Simon Glover, Prof. Dr. Francesco Palla, Dr. Raffaella Schneider and Prof. Dr. Marco Spaans, with whom I have published some of the works presented here in scientific journals. I further thank all members of the Institute of Theoretical Astrophysics for the very fruitful atmosphere and the stimulating discussions, as well as my parents and my sister Carolin for their support during my studies.

Bibliography

Abel, T., Anninos, P., Zhang, Y., & Norman, M. L. 1997, New Astronomy, 2, 181 23, 76, 187

Abel, T., Bryan, G. L., & Norman, M. L. 2002, Science, 295, 93 ix, 5, 19, 20, 28, 46, 98, 168, 176

Achterberg, A., Ackermann, M., Adams, J., Ahrens, J., Andeen, K., Auffenberg, J., Bai, X., Baret, B., Barwick, S. W., Bay, R., Beattie, K., Becka, T., Becker, J. K., Becker, K.-H., Beimforde, M., Berghaus, P., Berley, D., Bernardini, E., Bertrand, D., Besson, D. Z., Blaufuss, E., Boersma, D. J., Bohm, C., Bolmont, J., Böser, S., Botner, O., Bouchta, A., Braun, J., Burgess, C., Burgess, T., Castermans, T., Chirkin, D., Christy, B., Clem, J., Cowen, D. F., D'Agostino, M. V., Davour, A., Day, C. T., de Clercq, C., Demirörs, L., Descamps, F., Desiati, P., Deyoung, T., Diaz-Velez, J. C., Dreyer, J., Dumm, J. P., Duvoort, M. R., Edwards, W. R., Ehrlich, R., Eisch, J., Ellsworth, R. W., Evenson, P. A., Fadiran, O., Fazely, A. R., Filimonov, K., Finley, C., Foerster, M. M., Fox, B. D., Franckowiak, A., Franke, R., Gaisser, T. K., Gallagher, J., Ganugapati, R., Geenen, H., Gerhardt, L., Goldschmidt, A., Goodman, J. A., Gozzini, R., Griesel, T., Grullon, S., Groß, A., Gunasingha, R. M., Gurtner, M., Ha, C., Hallgren, A., Halzen, F., Han, K., Hanson, K., Hardtke, D., Hardtke, R., Hart, J. E., Hasegawa, Y., Hauschildt, T., Hays, D., Heise, J., Helbing, K., Hellwig, M., Herquet, P., Hill, G. C., Hodges, J., Hoffman, K. D., Hommez, B., Hoshina, K., Hubert, D., Hughey, B., Hülß, J.-P., Hulth, P. O., Hultqvist, K., Hundertmark, S., Inaba, M., Ishihara, A., Jacobsen, J., Japaridze, G. S., Johansson, H., Jones, A., Joseph, J. M., Kampert, K.-H., Kappes, A., Karg, T., Karle, A., Kawai, H., Kelley, J. L., Kislat, F., Kitamura, N., Klein, S. R., Klepser, S., Kohnen, G., Kolanoski, H., Köpke, L., Kowalski, M., Kowarik, T., Krasberg, M., Kuehn, K., Labare, M., Landsman, H., Lauer, R., Leich, H., Leier, D., Liubarsky, I., Lundberg, J., Lünemann, J., Madsen, J., Maruyama, R., Mase, K., Matis, H. S., McCauley, T., McParland, C. P., Meagher, K., Meli, A., Messarius, T., Mészáros, P., Miyamoto, H., Mokhtarani, A., Montaruli, T., Morey, A., Morse, R., Movit, S. M., Münich, K., Nahnhauer, R., Nam, J. W., Nießen, P., Nygren, D. R., Olivas, A., Patton, S., Peña-Garay, C., de Los Heros, C. P., Piegsa, A., Pieloth, D., Pohl, A. C., Porrata, R., Pretz, J., Price, P. B., Przybylski,

BIBLIOGRAPHY

G. T., Rawlins, K., Razzaque, S., Redl, P., Resconi, E., Rhode, W., Ribordy, M., Rizzo, A., Robbins, S., Roth, P., Rothmaier, F., Rott, C., Rutledge, D., Ryckbosch, D., Sander, H.-G., Sarkar, S., Satalecka, K., Schlenstedt, S., Schmidt, T., Schneider, D., Seckel, D., Semburg, B., Seo, S. H., Sestayo, Y., Seunarine, S., Silvestri, A., Smith, A. J., Song, C., Sopher, J. E., Spiczak, G. M., Spiering, C., Stamatikos, M., Stanev, T., Stezelberger, T., Stokstad, R. G., Stoufer, M. C., Stoyanov, S., Strahler, E. A., Straszheim, T., Sulanke, K.-H., Sullivan, G. W., Sumner, T. J., Taboada, I., Tarasova, O., Tepe, A., Thollander, L., Tilav, S., Tluczykont, M., Toale, P. A., Tosi, D., Turčan, D., van Eijndhoven, N., Vandenbroucke, J., van Overloop, A., de Vries-Uiterweerd, G., Viscomi, V., Voigt, B., Wagner, W., Walck, C., Waldmann, H., Walter, M., Wang, Y.-R., Wendt, C., Wiebusch, C. H., Wikström, G., Williams, D. R., Wischnewski, R., Wissing, H., Woschnagg, K., Xu, X. W., Yodh, G., Yoshida, S., & Zornoza, J. D. 2007, Phys. Rev. D, 76, 027101 113

Adams, J., Danielsson, U. H., Grasso, D., & Rubinstein, H. 1996, Physics Letters B, 388, 253 21

Ahmed, Z., Akerib, D. S., Attisha, M. J., Bailey, C. N., Baudis, L., Bauer, D. A., Brink, P. L., Brusov, P. P., Bunker, R., Cabrera, B., Caldwell, D. O., Chang, C. L., Cooley, J., Crisler, M. B., Cushman, P., Daal, M., Dejongh, F., Dixon, R., Dragowsky, M. R., Duong, L., Ferril, R., Figueroa-Feliciano, E., Filippini, J., Gaitskell, R. J., Golwala, S. R., Grant, D. R., Hennings-Yeomans, R., Holmgren, D., Huber, M. E., Kamat, S., Leclercq, S., Mahapatra, R., Mandic, V., Meunier, P., Mirabolfathi, N., Nelson, H., Ogburn, R. W., Pyle, M., Qiu, X., Ramberg, E., Rau, W., Reisetter, A., Ross, R. R., Saab, T., Sadoulet, B., Sander, J., Schnee, R. W., Seitz, D. N., Serfass, B., Sundqvist, K. M., Thompson, J.-P. F., Wang, G., Yellin, S., Yoo, J., & Young, B. A. 2008, Journal of Low Temperature Physics, 151, 800 14

Ahn, K., & Komatsu, E. 2005a, Phys. Rev. D, 71, 021303 112, 117, 119, 120, 128, 130

—. 2005b, Phys. Rev. D, 72, 061301 14, 112, 117, 119, 120, 122, 128, 129, 130

Ahn, K., Komatsu, E., & Höflich, P. 2005, Phys. Rev. D, 71, 121301 14, 98, 117, 128

Ahrens, J., Andrés, E., Bai, X., Barouch, G., Barwick, S. W., Bay, R. C., Becka, T., Becker, K.-H., Bertrand, D., Biron, A., Botner, O., Bouchta, A., Carius, S., Chen, A., Chirkin, D., Conrad, J., Cooley, J., Costa, C. G., Cowen, D. F., Dalberg, E., de Clercq, C., Deyoung, T., Desiati, P., Dewulf, J.-P., Doksus, P., Edsjö, J., Ekström, P., Feser, T., Gaisser, T. K., Gaug, M., Gerhardt, L., Goldschmidt, A., Goobar, A., Hallgren, A., Halzen, F., Hanson, K., Hardtke, R., Hauschildt, T., Hellwig, M., Hill, G. C., Hulth, P. O., Hundertmark, S., Jacobsen, J., Karle, A., Kim, J., Koci, B., Köpke, L., Kowalski, M., Lamoureux, J. I., Leich, H., Leuthold, M., Lindahl, P., Loaiza, P., Lowder, D. M., Ludvig, J., Madsen, J., Marciniewski, P., Matis, H. S., McParland, C. P., Miller, T. C., Minaeva, Y., Miočinović, P., Mock, P. C., Morse, R., Neunhöffer, T., Niessen, P., Nygren, D. R., Ogelman, H.,

BIBLIOGRAPHY

Olbrechts, P., Pérez de Los Heros, C., Pohl, A., Porrata, R., Price, P. B., Przybylski, G. T., Rawlins, K., Rhode, W., Ribordy, M., Richter, S., Rodríguez Martino, J., Romenesko, P., Ross, D., Sander, H.-G., Schmidt, T., Schneider, D., Schneider, E., Schwarz, R., Silvestri, A., Solarz, M., Spiczak, G. M., Spiering, C., Steele, D., Steffen, P., Stokstad, R. G., Streicher, O., Sudhoff, P., Sulanke, K. H., Taboada, I., Thollander, L., Thon, T., Tilav, S., Vander Donckt, M., Walck, C., Weinheimer, C., Wiebusch, C. H., Wiedemann, C., Wischnewski, R., Wissing, H., Woschnagg, K., Wu, W., Yodh, G., & Young, S. 2002, Phys. Rev. D, 66, 032006 113

Ajello, M., Greiner, J., Sato, G., Willis, D. R., Kanbach, G., Strong, A. W., Diehl, R., Hasinger, G., Gehrels, N., Markwardt, C. B., & Tueller, J. 2008, ArXiv 0808.3377 xii, xiii, 118, 120, 121, 122, 123

Allison, A. C., & Dalgarno, A. 1969, ApJ, 158, 423 46

Alpher, R. A., Bethe, H., & Gamow, G. 1948, Physical Review, 73, 803 4

Ando, S. 2005, Physical Review Letters, 94, 171303 112, 119, 128, 129

Ando, S., Komatsu, E., Narumoto, T., & Totani, T. 2007, Phys. Rev. D, 75, 063519 98, 133

Angle, J., Aprile, E., Arneodo, F., Baudis, L., Bernstein, A., Bolozdynya, A., Coelho, L. C. C., Dahl, C. E., Deviveiros, L., Ferella, A. D., Fernandes, L. M. P., Fiorucci, S., Gaitskell, R. J., Giboni, K. L., Gomez, R., Hasty, R., Kastens, L., Kwong, J., Lopes, J. A. M., Madden, N., Manalaysay, A., Manzur, A., McKinsey, D. N., Monzani, M. E., Ni, K., Oberlack, U., Orboeck, J., Plante, G., Santorelli, R., Dos Santos, J. M. F., Shagin, P., Shutt, T., Sorensen, P., Schulte, S., Winant, C., & Yamashita, M. 2008, Physical Review Letters, 101, 091301 14

Anninos, P., & Norman, M. L. 1996, ApJ, 460, 556 74, 82, 83

Anninos, P., Zhang, Y., Abel, T., & Norman, M. L. 1997, New Astronomy, 2, 209 23, 68, 69, 74, 77, 78, 83

Ashie, Y., Hosaka, J., Ishihara, K., Itow, Y., Kameda, J., Koshio, Y., Minamino, A., Mitsuda, C., Miura, M., Moriyama, S., Nakahata, M., Namba, T., Nambu, R., Obayashi, Y., Shiozawa, M., Suzuki, Y., Takeuchi, Y., Taki, K., Yamada, S., Ishitsuka, M., Kajita, T., Kaneyuki, K., Nakayama, S., Okada, A., Okumura, K., Saji, C., Takenaga, Y., Clark, S. T., Desai, S., Kearns, E., Likhoded, S., Stone, J. L., Sulak, L. R., Wang, W., Goldhaber, M., Casper, D., Cravens, J. P., Gajewski, W., Kropp, W. R., Liu, D. W., Mine, S., Smy, M. B., Sobel, H. W., Sterner, C. W., Vagins, M. R., Ganezer, K. S., Hill, J., Keig, W. E., Jang, J. S., Kim, J. Y., Lim, I. T., Scholberg, K., Walter, C. W., Ellsworth, R. W., Tasaka, S., Guillian, G., Kibayashi, A., Learned, J. G., Matsuno, S., Takemori, D., Messier, M. D., Hayato, Y., Ichikawa, A. K., Ishida, T., Ishii, T., Iwashita, T., Kobayashi, T., Maruyama, T., Nakamura, K., Nitta, K., Oyama, Y., Sakuda, M., Totsuka, Y., Suzuki, A. T., Hasegawa,

BIBLIOGRAPHY

M., Hayashi, K., Kato, I., Maesaka, H., Morita, T., Nakaya, T., Nishikawa, K., Sasaki, T., Ueda, S., Yamamoto, S., Haines, T. J., Dazeley, S., Hatakeyama, S., Svoboda, R., Blaufuss, E., Goodman, J. A., Sullivan, G. W., Turcan, D., Habig, A., Fukuda, Y., Jung, C. K., Kato, T., Kobayashi, K., Malek, M., Mauger, C., McGrew, C., Sarrat, A., Sharkey, E., Yanagisawa, C., Toshito, T., Miyano, K., Tamura, N., Ishii, J., Kuno, Y., Yoshida, M., Kim, S. B., Yoo, J., Okazawa, H., Ishizuka, T., Choi, Y., Seo, H. K., Gando, Y., Hasegawa, T., Inoue, K., Shirai, J., Suzuki, A., Koshiba, M., Nakajima, Y., Nishijima, K., Harada, T., Ishino, H., Watanabe, Y., Kielczewska, D., Zalipska, J., Berns, H. G., Gran, R., Shiraishi, K. K., Stachyra, A., Washburn, K., & Wilkes, R. J. 2005, Phys. Rev. D, 71, 112005 113

Avelino, P. P., & Barbosa, D. 2004, Phys. Rev. D, 70, 067302 22

Bamba, K., Ohta, N., & Tsujikawa, S. 2008, Phys. Rev. D, 78, 043524 13

Banerjee, R., & Jedamzik, K. 2003, Physical Review Letters, 91, 251301 9, 31, 32, 51, 172

—. 2004, Phys. Rev. D, 70, 123003 32

Banerjee, R., Pudritz, R. E., & Holmes, L. 2004, MNRAS, 355, 248 12, 51

Barkana, R., & Loeb, A. 2001, Phys. Rep., 349, 125 7, 20, 24, 27, 50, 101

—. 2005a, ApJ, 626, 1 46, 56, 57, 59

—. 2005b, ApJ, 626, 1 60, 108

—. 2005c, MNRAS, 363, L36 60

Barrow, J. D., Ferreira, P. G., & Silk, J. 1997, Physical Review Letters, 78, 3610 21, 43

Basu, K. 2007, New Astronomy Review, 51, 431 71, 95

Basu, K., Hernández-Monteagudo, C., & Sunyaev, R. A. 2004, A&A, 416, 447 43, 71, 89, 94, 95

Baym, G., Bödeker, D., & McLerran, L. 1996, Phys. Rev. D, 53, 662 11

Beacom, J. F., Bell, N. F., & Bertone, G. 2005, Physical Review Letters, 94, 171301 14, 117, 128

Beacom, J. F., Bell, N. F., & Mack, G. D. 2007, Physical Review Letters, 99, 231301 110, 113

Beacom, J. F., & Yüksel, H. 2006, Physical Review Letters, 97, 071102 xiii, 14, 128, 131, 132

Bean, R., Melchiorri, A., & Silk, J. 2003, Phys. Rev. D, 68, 083501 22

Beck, R. 2001, Space Science Reviews, 99, 243 21

Beck, R., Brandenburg, A., Moss, D., Shukurov, A., & Sokoloff, D. 1996, ARA&A, 34, 155 21

BIBLIOGRAPHY

Becker, R. H., Fan, X., White, R. L., Strauss, M. A., Narayanan, V. K., Lupton, R. H., Gunn, J. E., Annis, J., Bahcall, N. A., Brinkmann, J., Connolly, A. J., Csabai, I., Czarapata, P. C., Doi, M., Heckman, T. M., Hennessy, G. S., Ivezić, Ž., Knapp, G. R., Lamb, D. Q., McKay, T. A., Munn, J. A., Nash, T., Nichol, R., Pier, J. R., Richards, G. T., Schneider, D. P., Stoughton, C., Szalay, A. S., Thakar, A. R., & York, D. G. 2001, AJ, 122, 2850 7, 20, 46, 99, 103, 105

Begelman, M. C., Volonteri, M., & Rees, M. J. 2006, MNRAS, 370, 289 6, 152

Bell, K. L., & Berrington, K. A. 1987, Journal of Physics B Atomic Molecular Physics, 20, L353 86, 179, 181

Berezhiani, Z., & Dolgov, A. D. 2004, Astroparticle Physics, 21, 59 10

Berezhiani, Z. G., Khlopov, M. Y., & Khomeriki, R. R. 1990, Yadernaya Fizika, 52, 104 22

Bergström, L. 2000, Reports on Progress in Physics, 63, 793 14

Bernet, M. L., Miniati, F., Lilly, S. J., Kronberg, P. P., & Dessauges-Zavadsky, M. 2008, Nature, 454, 302 9, 21

Bersanelli, M., Bouchet, F. R., Efstathiou, G., Griffin, M., Lamarre, J. M., Mandolesi, N., Norgaard-Nielsen, H. U., Pace, O., Polny, J., Puget, J. L., Tauber, J., Vittorio, N., & Volonté, S. 1996, COBRAS/SAMBA. A mission dedicated to imaging the anisotropies of the cosmic microwave background. Report on the phase A study. (COBRAS/SAMBA. A mission dedicated to imaging the anisotropies of the cosmic microwave background. Report on the phase A study., by Bersanelli, M.; Bouchet, F. R.; Efstathiou, G.; Griffin, M.; Lamarre, J. M.; Mandolesi, N.; Norgaard-Nielsen, H. U.; Pace, O.; Polny, J.; Puget, J. L.; Tauber, J.; Vittorio, N.; Volonté, S.. European Space Agency, Paris (France), Feb 1996, XII + 115,) 95

Bertone, G., Hooper, D., & Silk, J. 2004, Phys. Rep., 405, 279 14

Bharadwaj, S., & Ali, S. S. 2004, MNRAS, 352, 142 60

Biermann, L. 1950, Zeitschrift Naturforschung Teil A, 5, 65 174

Black, J. H. 2006, Faraday Discuss., 133, 27 67, 69, 83, 86, 90, 170

Bland-Hawthorn, J., & Maloney, P. R. 1999, ApJ, 510, L33 29

—. 2001, ApJ, 550, L231 29

Bœhm, C., & Fayet, P. 2004, Nuclear Physics B, 683, 219 133

Boehm, C., Hooper, D., Silk, J., Casse, M., & Paul, J. 2004a, Physical Review Letters, 92, 101301 14, 97, 116, 117, 127, 128

—. 2004b, Physical Review Letters, 92, 101301 116, 128, 170

Bonometto, S. A., & Pantano, O. 1993, Phys. Rep., 228, 175 3

BIBLIOGRAPHY

Bougleux, E., & Galli, D. 1997, MNRAS, 288, 638 84

Bradford, C. M., Aguirre, J. E., Aikin, R., Bock, J. J., Earle, L., Glenn, J., Inami, H., Maloney, P. R., Matsuhara, H., Naylor, B. J., Nguyen, H. T., & Zmuidzinas, J. 2009, ApJ, 705, 112 136, 147

Brage, T., & Fischer, C. F. 1991, Phys. Rev. A, 44, 72 88

Brandenburg, A., & Subramanian, K. 2005, Phys. Rep., 417, 1 175

Broadhurst, T., Benítez, N., Coe, D., Sharon, K., Zekser, K., White, R., Ford, H., Bouwens, R., Blakeslee, J., Clampin, M., Cross, N., Franx, M., Frye, B., Hartig, G., Illingworth, G., Infante, L., Menanteau, F., Meurer, G., Postman, M., Ardila, D. R., Bartko, F., Brown, R. A., Burrows, C. J., Cheng, E. S., Feldman, P. D., Golimowski, D. A., Goto, T., Gronwall, C., Herranz, D., Holden, B., Homeier, N., Krist, J. E., Lesser, M. P., Martel, A. R., Miley, G. K., Rosati, P., Sirianni, M., Sparks, W. B., Steindling, S., Tran, H. D., Tsvetanov, Z. I., & Zheng, W. 2005, ApJ, 621, 53 129

Bromm, V., Ferrara, A., Coppi, P. S., & Larson, R. B. 2001, MNRAS, 328, 969 5, 6, 29, 36, 100, 173

Bromm, V., & Larson, R. B. 2004, ARA&A, 42, 79 5, 19, 20, 28, 46, 98, 168

Bryan, G. L., Norman, M. L., Stone, J. M., Cen, R., & Ostriker, J. P. 1995, Computer Physics Communications, 89, 149 69, 77

Buhr, H. e. a. 2007, Phys. Rev. A, submitted 76

Burkert, A. 1995, ApJ, 447, L25+ 128, 129

Campanelli, L. 2008, ArXiv 0805.0575 13

Campanelli, L., Cea, P., Fogli, G. L., & Tedesco, L. 2008, Phys. Rev. D, 77, 043001 13

Capitelli, M., Coppola, C. M., Diomede, P., & Longo, S. 2007, A&A, 470, 811 76, 187

Carilli, C. L., Cox, P., Bertoldi, F., Menten, K. M., Omont, A., Djorgovski, S. G., Petric, A., Beelen, A., Isaak, K. G., & McMahon, R. G. 2002, ApJ, 575, 145 21, 135, 140

Chabrier, G. 2003, PASP, 115, 763 19, 29, 65

Chartas, G., Brandt, W. N., Gallagher, S. C., & Garmire, G. P. 2002, ApJ, 579, 169 149

Chen, X., & Kamionkowski, M. 2004, Phys. Rev. D, 70, 043502 22

Chen, X., & Miralda-Escudé, J. 2004, ApJ, 602, 1 57, 59

Cheng, B., & Olinto, A. V. 1994, Phys. Rev. D, 50, 2421 11

Chluba, J., & Sunyaev, R. A. 2006, A&A, 446, 39 68, 94

Choudhury, T. R., & Ferrara, A. 2005, MNRAS, 361, 577 7, 20, 24, 50, 101

—. 2007, MNRAS, 380, L6 153

Christensson, M., Hindmarsh, M., & Brandenburg, A. 2001, Phys. Rev. E, 64, 056405 31, 32, 51

Chuzhoy, L. 2008, ApJ, 679, L65 41, 45, 112, 119, 128, 130, 133, 169

Chuzhoy, L., Alvarez, M. A., & Shapiro, P. R. 2006, ApJ, 648, L1 57

Chuzhoy, L., & Shapiro, P. R. 2006, ApJ, 651, 1 56, 59

Ciardi, B., Bianchi, S., & Ferrara, A. 2002, MNRAS, 331, 463 28, 59, 104

Ciardi, B., Ferrara, A., & White, S. D. M. 2003, MNRAS, 344, L7 7, 20

Cirelli, M., Franceschini, R., & Strumia, A. 2008, Nuclear Physics B, 800, 204 14, 97, 127

Clark, P. C., Glover, S. C. O., & Klessen, R. S. 2008, ApJ, 672, 757 6, 20, 28, 36, 46, 104, 107, 173

Colbourn, E. A., & Bunker, P. R. 1976, Journal of Molecular Spectroscopy, 63, 155 85

Cowling, T. G. 1956, MNRAS, 116, 114 30, 50

Cumberbatch, D. T., Lattanzi, M., & Silk, J. 2008, ArXiv 0808.0881 41, 45, 112, 119, 128, 129, 131, 169

Dalgarno, A., & Lane, N. F. 1966, ApJ, 145, 623 xvi, 86, 179, 180

Dalgarno, A., Yan, M., & Liu, W. 1999, ApJS, 125, 237 181

Davies, R. I., Mueller Sánchez, F., Genzel, R., Tacconi, L. J., Hicks, E. K. S., Friedrich, S., & Sternberg, A. 2007, ApJ, 671, 1388 141

de Blok, W. J. G., & Bosma, A. 2002, A&A, 385, 816 129

de Boer, W. 2008, ArXiv 0810.1472 14, 98, 127

de Boer, W., Sander, C., Zhukov, V., Gladyshev, A. V., & Kazakov, D. I. 2005, A&A, 444, 51 14, 97, 127

Deharveng, J.-M., Buat, V., Le Brun, V., Milliard, B., Kunth, D., Shull, J. M., & Gry, C. 2001, A&A, 375, 805 29

Desai, S., Ashie, Y., Fukuda, S., Fukuda, Y., Ishihara, K., Itow, Y., Koshio, Y., Minamino, A., Miura, M., Moriyama, S., Nakahata, M., Namba, T., Nambu, R., Obayashi, Y., Sakurai, N., Shiozawa, M., Suzuki, Y., Takeuchi, H., Takeuchi, Y., Yamada, S., Ishitsuka, M., Kajita, T., Kaneyuki, K., Nakayama, S., Okada, A., Ooyabu, T., Saji, C., Earl, M., Kearns, E., Stone, J. L., Sulak, L. R., Walter, C. W., Wang, W., Goldhaber, M., Barszczak, T., Casper, D., Cravens, J. P., Gajewski, W., Kropp, W. R., Mine, S., Liu, D. W., Smy, M. B., Sobel, H. W., Sterner, C. W., Vagins, M. R., Ganezer, K. S., Hill, J., Keig, W. E., Kim, J. Y., Lim, I. T., Ellsworth, R. W., Tasaka, S., Guillian, G., Kibayashi, A., Learned, J. G., Matsuno, S.,

BIBLIOGRAPHY

Takemori, D., Messier, M. D., Hayato, Y., Ichikawa, A. K., Ishida, T., Ishii, T., Iwashita, T., Kameda, J., Kobayashi, T., Maruyama, T., Nakamura, K., Nitta, K., Oyama, Y., Sakuda, M., Totsuka, Y., Suzuki, A. T., Hasegawa, M., Hayashi, K., Inagaki, T., Kato, I., Maesaka, H., Morita, T., Nakaya, T., Nishikawa, K., Sasaki, T., Ueda, S., Yamamoto, S., Haines, T. J., Dazeley, S., Hatakeyama, S., Svoboda, R., Blaufuss, E., Goodman, J. A., Sullivan, G. W., Turcan, D., Scholberg, K., Habig, A., Jung, C. K., Kato, T., Kobayashi, K., Malek, M., Mauger, C., McGrew, C., Sarrat, A., Sharkey, E., Yanagisawa, C., Toshito, T., Mitsuda, C., Miyano, K., Shibata, T., Kajiyama, Y., Nagashima, Y., Takita, M., Yoshida, M., Kim, H. I., Kim, S. B., Yoo, J., Okazawa, H., Ishizuka, T., Choi, Y., Seo, H. K., Gando, Y., Hasegawa, T., Inoue, K., Shirai, J., Suzuki, A., Koshiba, M., Hashimoto, T., Nakajima, Y., Nishijima, K., Harada, T., Ishino, H., Morii, M., Nishimura, R., Watanabe, Y., Kielczewska, D., Zalipska, J., Gran, R., Shiraishi, K. K., Washburn, K., & Wilkes, R. J. 2004, Phys. Rev. D, 70, 109901 14

Di Matteo, T., Perna, R., Abel, T., & Rees, M. J. 2002, ApJ, 564, 576 65, 168

Dicke, R. H., Peebles, P. J. E., Roll, P. G., & Wilkinson, D. T. 1965, ApJ, 142, 414 4

Dijkstra, M., Haiman, Z., & Loeb, A. 2004, ApJ, 613, 646 7, 20, 110

Dijkstra, M., Haiman, Z., Mesinger, A., & Wyithe, S. 2008, ArXiv e-prints 6, 153

Dine, M., Leigh, R. G., Huet, P., Linde, A., & Linde, D. 1992, Phys. Rev. D, 46, 550 3

Dolag, K., Bartelmann, M., & Lesch, H. 1999, A&A, 348, 351 9

Donnert, J., Dolag, K., Lesch, H., & Müller, E. 2009, MNRAS, 392, 1008 9

Doroshkevich, A. G., Khlopov, M. I., & Klypin, A. A. 1989, MNRAS, 239, 923 22

Dove, J. B., Shull, J. M., & Maloney, P. R. 2000, in Bulletin of the American Astronomical Society, Vol. 32, Bulletin of the American Astronomical Society, 1467–+ 28, 104

Draine, B. T. 1980, ApJ, 241, 1021 30

Drees, M. 1996, ArXiv High Energy Physics - Phenomenology e-prints 3

Drees, M., & Nojiri, M. M. 1993, Phys. Rev. D, 47, 376 98, 117, 128

Dubrovich, V. K. 1975, Soviet Astronomy Letters, 1, 196 68

—. 1994, Astronomical and Astrophysical Transactions, 5, 57 69, 84, 85

—. 1997, A&A, 324, 27 69, 71, 89, 94

Dubrovich, V. K., Bajkova, A. T., & Khaikin, V. B. 2007, SSF as a Manifestation of Protoobjects in the Dark Ages Epoch: Theory and Experiment (Exploring the Cosmic Frontier: Astrophysical Instruments for the 21st Century), 109–+ 69

Einasto, J. 1965, Trudy Inst. Astroz. Alma-Ata, 51, 87 129

BIBLIOGRAPHY

Eisenstein, D. J., & Loeb, A. 1995, ApJ, 443, 11 6, 152

Ellis, J. 2000, Physica Scripta Volume T, 85, 221 14

Engel, E. A., Doss, N., Harris, G. J., & Tennyson, J. 2005, MNRAS, 357, 471 84, 85

Espinosa, J. R., Quirós, M., & Zwirner, F. 1993, Physics Letters B, 314, 206 3

Fan, X., Narayanan, V. K., Lupton, R. H., Strauss, M. A., Knapp, G. R., Becker, R. H., White, R. L., Pentericci, L., Leggett, S. K., Haiman, Z., Gunn, J. E., Ivezić, Ž., Schneider, D. P., Anderson, S. F., Brinkmann, J., Bahcall, N. A., Connolly, A. J., Csabai, I., Doi, M., Fukugita, M., Geballe, T., Grebel, E. K., Harbeck, D., Hennessy, G., Lamb, D. Q., Miknaitis, G., Munn, J. A., Nichol, R., Okamura, S., Pier, J. R., Prada, F., Richards, G. T., Szalay, A., & York, D. G. 2001, AJ, 122, 2833 1

Fan, X., Strauss, M. A., Richards, G. T., Hennawi, J. F., Becker, R. H., White, R. L., Diamond-Stanic, A. M., Donley, J. L., Jiang, L., Kim, J. S., Vestergaard, M., Young, J. E., Gunn, J. E., Lupton, R. H., Knapp, G. R., Schneider, D. P., Brandt, W. N., Bahcall, N. A., Barentine, J. C., Brinkmann, J., Brewington, H. J., Fukugita, M., Harvanek, M., Kleinman, S. J., Krzesinski, J., Long, D., Neilsen, Jr., E. H., Nitta, A., Snedden, S. A., & Voges, W. 2006, AJ, 131, 1203 152

Fayet, P., Hooper, D., & Sigl, G. 2006, Physical Review Letters, 96, 211302 15, 128

Feretti, L., Dallacasa, D., Giovannini, G., & Tagliani, A. 1995, A&A, 302, 680 9

Fernández-Soto, A., Lanzetta, K. M., & Chen, H.-W. 2003, MNRAS, 342, 1215 30, 105

Field, G. B. 1958, Proc. I.R.E., 46, 240 47, 95, 108

Finkbeiner, D. P., Padmanabhan, N., & Weiner, N. 2008, Phys. Rev. D, 78, 063530 15

Finkbeiner, D. P., & Weiner, N. 2007, Phys. Rev. D, 76, 083519 15

Fixsen, D. J., Cheng, E. S., Gales, J. M., Mather, J. C., Shafer, R. A., & Wright, E. L. 1996, ApJ, 473, 576 94

Fixsen, D. J., & Mather, J. C. 2002, ApJ, 581, 817 4, 74

Flores, R. A., & Primack, J. R. 1994, ApJ, 427, L1 129

Flower, D. R., Le Bourlot, J., Pineau des Forêts, G., & Roueff, E. 2000, MNRAS, 314, 753 82

Fowler, J. W. 2004, in Presented at the Society of Photo-Optical Instrumentation Engineers (SPIE) Conference, Vol. 5498, Society of Photo-Optical Instrumentation Engineers (SPIE) Conference Series, ed. C. M. Bradford, P. A. R. Ade, J. E. Aguirre, J. J. Bock, M. Dragovan, L. Duband, L. Earle, J. Glenn, H. Matsuhara, B. J. Naylor, H. T. Nguyen, M. Yun, & J. Zmuidzinas, 1–10 95

BIBLIOGRAPHY

Freese, K., Bodenheimer, P., Spolyar, D., & Gondolo, P. 2008a, ArXiv e-prints, 806 16, 20, 33, 35, 98, 99, 100, 110, 115, 123, 153, 168

Freese, K., Gondolo, P., Sellwood, J. A., & Spolyar, D. 2008b, ArXiv e-prints, 805 20, 33, 110

Freese, K., Spolyar, D., & Aguirre, A. 2008c, ArXiv e-prints, 802 16, 20, 33, 98, 99, 106, 110, 168

Fujita, A., Martin, C. L., Mac Low, M.-M., & Abel, T. 2003, ApJ, 599, 50 28, 104

Fukushige, T., & Makino, J. 1997, ApJ, 477, L9+ 129

—. 2003, ApJ, 588, 674 129

Furlanetto, S. R., & Furlanetto, M. R. 2007, MNRAS, 379, 130 46, 56

Furlanetto, S. R., Oh, S. P., & Briggs, F. H. 2006a, Phys. Rep., 433, 181 27, 28, 53, 56, 57, 59, 60, 64

Furlanetto, S. R., Oh, S. P., & Pierpaoli, E. 2006b, Phys. Rev. D, 74, 103502 22, 33, 34, 35, 45, 46, 60, 110, 117, 133, 169

Furlanetto, S. R., & Pritchard, J. R. 2006, MNRAS, 372, 1093 59

Gaisser, T. K., & Honda, M. 2002, Annual Review of Nuclear and Particle Science, 52, 153 113

Galli, D., & Palla, F. 1998, A&A, 335, 403 4, 68, 75, 76, 77, 81, 82, 83, 84, 101, 183, 187

—. 2002, Planet. Space Sci., 50, 1197 77, 187

Galliano, E., & Alloin, D. 2002, A&A, 393, 43 148

Galliano, E., Alloin, D., Granato, G. L., & Villar-Martín, M. 2003, A&A, 412, 615 xv, 136, 140, 141, 143, 148, 164

Gamgami, F. 2007, PhD thesis, University of Heidelberg 36

Gamow, G. 1946, Physical Review, 70, 572 4

Gentile, G., Salucci, P., & Klein, U. 2004, in Baryons in Dark Matter Halos, ed. R. Dettmar, U. Klein, & P. Salucci 129

Ghigna, S., Moore, B., Governato, F., Lake, G., Quinn, T., & Stadel, J. 2000, ApJ, 544, 616 129

Giallongo, E., Cristiani, S., D'Odorico, S., & Fontana, A. 2002, ApJ, 568, L9 29, 105

Gilfanov, M., Grimm, H.-J., & Sunyaev, R. 2004, MNRAS, 347, L57 27

Gingerich, O. 1961, ApJ, 134, 653 xvi, 86, 179, 180

Glover, S. C., Savin, D. W., & Jappsen, A.-K. 2006, ApJ, 640, 553 93

BIBLIOGRAPHY

Glover, S. C. O., & Abel, T. 2008, MNRAS, 388, 1627 82

Glover, S. C. O., & Brand, P. W. J. L. 2001, MNRAS, 321, 385 6

—. 2003, MNRAS, 340, 210 27

Glover, S. C. O., & Jappsen, A.-K. 2007, ApJ, 666, 1 76, 187

Glover, S. C. O., & Savin, D. W. 2009, MNRAS, 393, 911 175

Gnedin, N. Y. 2000, ApJ, 542, 535 7, 20, 21, 26, 46, 52, 102, 119

Gnedin, N. Y., & Hui, L. 1998, MNRAS, 296, 44 20, 26, 46, 52, 102, 119

Goldman, S. P. 1989, Phys. Rev. A, 40, 1185 24, 50, 73

Graham, A. W., Merritt, D., Moore, B., Diemand, J., & Terzić, B. 2006, AJ, 132, 2701 129

Grasso, D., & Rubinstein, H. R. 2001, Phys. Rep., 348, 163 8, 9, 11, 12, 21

Greenhill, L. J., Gwinn, C. R., Antonucci, R., & Barvainis, R. 1996, ApJ, 472, L21+ 148

Greenstein, G. 1969, Nature, 223, 938 10, 21

Greif, T. H., Glover, S. C. O., Bromm, V., & Klessen, R. S. 2009, MNRAS, 392, 1381 104, 173

Greif, T. H., Johnson, J. L., Bromm, V., & Klessen, R. S. 2007, ApJ, 670, 1 6

Greif, T. H., Johnson, J. L., Klessen, R. S., & Bromm, V. 2008, MNRAS, 387, 1021 6, 27, 29, 36, 52, 104, 112, 119, 173, 176

Greve, T. R., Papadopoulos, P. P., Gao, Y., & Radford, S. J. E. 2009, ApJ, 692, 1432 140, 177

Griest, K. E. 1987, PhD thesis, AA(California Univ., Santa Cruz.) 116

Grimm, H.-J., Gilfanov, M., & Sunyaev, R. 2003, MNRAS, 339, 793 27

Gruber, D. E., Matteson, J. L., Peterson, L. E., & Jung, G. V. 1999, ApJ, 520, 124 xii, xiii, 118, 120, 121, 122, 123

Guth, A. H. 1981, Phys. Rev. D, 23, 347 2

Guth, A. H., & Pi, S.-Y. 1982, Physical Review Letters, 49, 1110 2

Haiman, Z., Abel, T., & Rees, M. J. 2000, ApJ, 534, 11 6

Haiman, Z., & Loeb, A. 1997a, ApJ, 483, 21 7, 20, 24, 50

—. 1997b, ApJ, 483, 21 101

Han, J. L. 2008, Nuclear Physics B Proceedings Supplements, 175, 62 8

Hansen, S. H., & Haiman, Z. 2004, ApJ, 600, 26 22

BIBLIOGRAPHY

Hao, L., Wu, Y., Charmandaris, V., Spoon, H. W. W., Bernard-Salas, J., Devost, D., Lebouteiller, V., & Houck, J. R. 2009, ApJ, 704, 1159 147

Harrison, E. R. 1970, MNRAS, 147, 279 10

Hawking, S. W. 1982, Physics Letters B, 115, 295 2

Hayashi, C. 1950, Progress of Theoretical Physics, 5, 224 4

Heckman, T. M., Sembach, K. R., Meurer, G. R., Leitherer, C., Calzetti, D., & Martin, C. L. 2001, ApJ, 558, 56 29

Heger, A., & Woosley, S. E. 2002, ApJ, 567, 532 5

Hennebelle, P., & Fromang, S. 2008, A&A, 477, 9 174

Hernández-Monteagudo, C., Haiman, Z., Jimenez, R., & Verde, L. 2007a, ApJ, 660, L85 95

Hernández-Monteagudo, C., Haiman, Z., Verde, L., & Jimenez, R. 2008, ApJ, 672, 33 95

Hernández-Monteagudo, C., Rubiño-Martín, J. A., & Sunyaev, R. A. 2007b, MNRAS, 380, 1656 95

Hernández-Monteagudo, C., Verde, L., & Jimenez, R. 2006, ApJ, 653, 1 95

Hinshaw, G., Nolta, M. R., Bennett, C. L., Bean, R., Doré, O., Greason, M. R., Halpern, M., Hill, R. S., Jarosik, N., Kogut, A., Komatsu, E., Limon, M., Odegard, N., Meyer, S. S., Page, L., Peiris, H. V., Spergel, D. N., Tucker, G. S., Verde, L., Weiland, J. L., Wollack, E., & Wright, E. L. 2007, ApJS, 170, 288 67

Hirata, C. M. 2006, MNRAS, 367, 259 56, 57, 81, 108

Hirata, C. M., & Padmanabhan, N. 2006, MNRAS, 372, 1175 69, 83

Hirata, C. M., & Switzer, E. R. 2008, Phys. Rev. D, 77, 083007 68

Hoekstra, H., Yee, H. K. C., & Gladders, M. D. 2004, in IAU Symposium, Vol. 220, Dark Matter in Galaxies, ed. S. Ryder, D. Pisano, M. Walker, & K. Freeman, 439–+ 129

Hogan, C. J. 1983, Physical Review Letters, 51, 1488 11

Hollenbach, D. J., & Tielens, A. G. G. M. 1999, Reviews of Modern Physics, 71, 173 138, 147

Holøien, E., & Midtdal, J. 1955, Proceedings of the Physical Society A, 68, 815 88

Honda, M., Kajita, T., Kasahara, K., & Midorikawa, S. 2004, Phys. Rev. D, 70, 043008 xii, 113, 114

Hooper, D., Finkbeiner, D. P., & Dobler, G. 2007, Phys. Rev. D, 76, 083012 14, 97, 127

Hummer, D. G. 1994, MNRAS, 268, 109 24, 49, 73

Hurwitz, M., Jelinsky, P., & Dixon, W. V. D. 1997, ApJ, 481, L31+ 29

BIBLIOGRAPHY

Inoue, A. K., Iwata, I., Deharveng, J.-M., Buat, V., & Burgarella, D. 2005, A&A, 435, 471 30, 105

Inoue, Y., Totani, T., & Ueda, Y. 2008, ApJ, 672, L5 98, 133

Iocco, F. 2008, ApJ, 677, L1 16, 19, 20, 33, 35, 98, 99, 110, 114, 116, 153

Iocco, F., Bressan, A., Ripamonti, E., Schneider, R., Ferrara, A., & Marigo, P. 2008, ArXiv 0805.4016 16, 20, 33, 98, 99, 100, 110, 112, 115, 116, 123, 153, 168

Jappsen, A. ., Mac Low, M. ., Glover, S. C. O., Klessen, R. S., & Kitsionas, S. 2008, ArXiv 0810.1867 6, 7

Jean, P., Knödlseder, J., Gillard, W., Guessoum, N., Ferrière, K., Marcowith, A., Lonjou, V., & Roques, J. P. 2006, A&A, 445, 579 14, 97, 127

Jedamzik, K., Katalinić, V., & Olinto, A. V. 1998, Phys. Rev. D, 57, 3264 31, 50

Jing, Y. P., & Suto, Y. 2000, ApJ, 529, L69 129

—. 2002, ApJ, 574, 538 129

John, T. L. 1988, A&A, 193, 189 xvi, 86, 87, 179, 180, 181

—. 1994, MNRAS, 269, 871 88

Johnson, J. L., & Bromm, V. 2006, MNRAS, 366, 247 104

Johnson, J. L., Greif, T. H., & Bromm, V. 2007a, ApJ, 665, 85 6, 65, 103

—. 2007b, ApJ, 665, 85 65, 103

—. 2008, MNRAS, 388, 26 6

Jones, B. J. T., & Wyse, R. F. G. 1985, A&A, 149, 144 68, 72

Juarez, Y., Maiolino, R., Mujica, R., Pedani, M., Marinoni, S., Nagao, T., Marconi, A., & Oliva, E. 2009, ArXiv e-prints 7, 9, 171

Juřek, M., Špirko, V., & Kraemer, W. P. 1995, Chem. Phys., 193, 287 187

Kajantie, K., & Kurki-Suonio, H. 1986, Phys. Rev. D, 34, 1719 3

Kajantie, K., Laine, M., Rummukainen, K., & Shaposhnikov, M. 1996, Physical Review Letters, 77, 2887 3

Kandus, A., Opher, R., & Barros, S. R. B. 2004, Braz. J. Phys., 34, 4b 6, 174

Kappadath, S. C., Ryan, J., Bennett, K., Bloemen, H., Forrest, D., Hermsen, W., Kippen, R. M., McConnell, M., Schoenfelder, V., van Dijk, R., Varendorff, M., Weidenspointner, G., & Winkler, C. 1996, A&AS, 120, C619+ xii, xiii, 121, 122, 131

Karr, J. P., & Hilico, L. 2006, Journal of Physics B Atomic Molecular Physics, 39, 2095 85

Kasuya, S., & Kawasaki, M. 2004, Phys. Rev. D, 70, 103519 22

BIBLIOGRAPHY

Kawakatu, N., & Wada, K. 2009, ApJ, 706, 676 153, 177

Kazantsev, A. P. 1968, Sov. Phys. JETP, 26, 1031 175

Kim, E.-J., Olinto, A. V., & Rosner, R. 1996, ApJ, 468, 28 21, 46, 64

Kim, K.-T., Kronberg, P. P., & Tribble, P. C. 1991, ApJ, 379, 80 9

Kinney, W. H. 2003, ArXiv astro-ph/0301448 2

Kinzer, R. L., Milne, P. A., Kurfess, J. D., Strickman, M. S., Johnson, W. N., & Purcell, W. R. 2001, ApJ, 559, 282 117, 119, 130

Klamer, I. J., Ekers, R. D., Sadler, E. M., Weiss, A., Hunstead, R. W., & De Breuck, C. 2005, ApJ, 621, L1 135, 140

Klypin, A., Kravtsov, A. V., Bullock, J. S., & Primack, J. R. 2001, ApJ, 554, 903 129

Knödlseder, J., Lonjou, V., Jean, P., Allain, M., Mandrou, P., Roques, J.-P., Skinner, G. K., Vedrenne, G., von Ballmoos, P., Weidenspointner, G., Caraveo, P., Cordier, B., Schönfelder, V., & Teegarden, B. J. 2003, A&A, 411, L457 14, 116

Kogut, A. 2003, New Astronomy Review, 47, 977 7

Kohler, K., Gnedin, N. Y., Miralda-Escudé, J., & Shaver, P. A. 2005, in Astronomical Society of the Pacific Conference Series, Vol. 345, Astronomical Society of the Pacific Conference Series, ed. N. Kassim, M. Perez, W. Junor, & P. Henning, 304–+ 7

Kolb, E. W., & Turner, M. S. 1990, The early universe (Frontiers in Physics, Reading, MA: Addison-Wesley, 1988, 1990) 98

Kollmeier, J. A., Onken, C. A., Kochanek, C. S., Gould, A., Weinberg, D. H., Dietrich, M., Cool, R., Dey, A., Eisenstein, D. J., Jannuzi, B. T., Le Floc'h, E., & Stern, D. 2006, ApJ, 648, 128 139

Komatsu, E., Dunkley, J., Nolta, M. R., Bennett, C. L., Gold, B., Hinshaw, G., Jarosik, N., Larson, D., Limon, M., Page, L., Spergel, D. N., Halpern, M., Hill, R. S., Kogut, A., Meyer, S. S., Tucker, G. S., Weiland, J. L., Wollack, E., & Wright, E. L. 2008, ArXiv e-prints, 803 1, 2, 20, 30, 33, 46, 68, 82, 99, 103, 130

Koushiappas, S. M., Bullock, J. S., & Dekel, A. 2004, MNRAS, 354, 292 6, 152

Koyama, K., Petre, R., Gotthelf, E. V., Hwang, U., Matsuura, M., Ozaki, M., & Holt, S. S. 1995, Nature, 378, 255 27, 58

Krause, F., & Rädler, K.-H. 1980, Mean-field magnetohydrodynamics and dynamo theory (Oxford, Pergamon Press, Ltd., 1980. 271 p.) 8

Kravtsov, A. V., Klypin, A. A., Bullock, J. S., & Primack, J. R. 1998, ApJ, 502, 48 129

Kronberg, P. P. 1994, Reports on Progress in Physics, 57, 325 8, 9

BIBLIOGRAPHY

Kroupa, P. 2002, Science, 295, 82 19, 29, 65

Kulsrud, R., Cowley, S. C., Gruzinov, A. V., & Sudan, R. N. 1997, Phys. Rep., 283, 213 8

Kulsrud, R. M., & Zweibel, E. G. 2008, Reports on Progress in Physics, 71, 046901 8

Kurki-Suonio, H. 1988, Phys. Rev. D, 37, 2104 11

Launay, J. M., Le Dourneuf, M., & Zeippen, C. J. 1991, A&A, 252, 842 94

Leitherer, C., Schaerer, D., Goldader, J. D., Delgado, R. M. G., Robert, C., Kune, D. F., de Mello, D. F., Devost, D., & Heckman, T. M. 1999, ApJS, 123, 3 65

Lepp, S., & Dalgarno, A. 1996, A&A, 306, L21 136, 138, 147

Lepp, S., Stancil, P. C., & Dalgarno, A. 2002, Journal of Physics B Atomic Molecular Physics, 35, 57 187

Linder, F., Janev, R. K., & Botero, J. 1995, in Atomic and Molecular Processes in Fusion Edge Plasmas, ed. R. K. Janev, 397–+ 187

Lipovka, A., Núñez-López, R., & Avila-Reese, V. 2005, MNRAS, 361, 850 82

Lodato, G., & Natarajan, P. 2006, MNRAS, 371, 1813 6, 152

Loeb, A., & Barkana, R. 2001, ARA&A, 39, 19 7, 20, 24, 50, 101

Loeb, A., & Zaldarriaga, M. 2004, Physical Review Letters, 92, 211301 45, 59

Loenen, A. F., Spaans, M., Baan, W. A., & Meijerink, R. 2008, A&A, 488, L5 137

Machacek, M. E., Bryan, G. L., & Abel, T. 2001, ApJ, 548, 509 6, 102

Machida, M. N., Omukai, K., Matsumoto, T., & Inutsuka, S.-i. 2006, ApJ, 647, L1 15, 46, 174, 175

Mack, G. D., Jacques, T. D., Beacom, J. F., Bell, N. F., & Yuksel, H. 2008, ArXiv 0803.0157 14, 42, 110, 111, 113, 117, 128, 129, 133, 170

Mackey, J., Bromm, V., & Hernquist, L. 2003, ApJ, 586, 1 27

Madau, P., Meiksin, A., & Rees, M. J. 1997, ApJ, 475, 429 59

Maiolino, R., Neri, R., Beelen, A., Bertoldi, F., Carilli, C. L., Caselli, P., Cox, P., Menten, K. M., Nagao, T., Omont, A., Walmsley, C. M., Walter, F., & Weiß, A. 2007, A&A, 472, L33 135, 140

Malkan, M., Webb, W., & Konopacky, Q. 2003, ApJ, 598, 878 30, 105

Maloney, P. R., Hollenbach, D. J., & Tielens, A. G. G. M. 1996, ApJ, 466, 561 136, 138, 147, 148

Maoli, R., Chambaud, P., Daniel, J. Y., de Bernardis, P., Encrenaz, P., Masi, S., Melchiorri,

BIBLIOGRAPHY

B., Melchiorri, F., Pagani, L., Rosmus, P., & Signore, M. 2005, in ESA Special Publication, Vol. 577, ESA Special Publication, ed. A. Wilson, 293–296 69

Maoli, R., Melchiorri, F., & Tosti, D. 1994, ApJ, 425, 372 67, 68, 71, 84

Mapelli, M., & Ferrara, A. 2005, MNRAS, 364, 2 117

Mapelli, M., Ferrara, A., & Pierpaoli, E. 2006, MNRAS, 369, 1719 117

Mapelli, M., & Ripamonti, E. 2007, Memorie della Societa Astronomica Italiana, 78, 800 103

Martin, P. G., Schwarz, D. H., & Mandy, M. E. 1996, ApJ, 461, 265 187

Matese, J. J., & O'Connell, R. F. 1969, Physical Review, 180, 1289 9, 21

—. 1970, ApJ, 160, 451 10

Mather, J. C., Cheng, E. S., Cottingham, D. A., Eplee, Jr., R. E., Fixsen, D. J., Hewagama, T., Isaacman, R. B., Jensen, K. A., Meyer, S. S., Noerdlinger, P. D., Read, S. M., Rosen, L. P., Shafer, R. A., Wright, E. L., Bennett, C. L., Boggess, N. W., Hauser, M. G., Kelsall, T., Moseley, Jr., S. H., Silverberg, R. F., Smoot, G. F., Weiss, R., & Wilkinson, D. T. 1994, ApJ, 420, 439 94

Matsuda, T., Sato, H., & Takeda, H. 1971, Progress of Theoretical Physics, 46, 416 68

Mayer, M., & Duschl, W. J. 2005, MNRAS, 358, 614 69

McCall, B. J., Geballe, T. R., Hinkle, K. H., & Oka, T. 1999, ApJ, 522, 338 140

Meijerink, R., & Spaans, M. 2005, A&A, 436, 397 136, 138, 139, 142, 171, 172, 177

Meijerink, R., Spaans, M., & Israel, F. P. 2006, ApJ, 650, L103 150

—. 2007, A&A, 461, 793 136, 143, 144, 146, 149, 155, 177

Mellema, G., Iliev, I. T., Pen, U.-L., & Shapiro, P. R. 2006, MNRAS, 372, 679 20, 24, 102

Miele, G., & Pisanti, O. 2008, ArXiv 0811.4479 4

Miralda-Escudé, J. 2003, ApJ, 597, 66 25

Moore, B., Quinn, T., Governato, F., Stadel, J., & Lake, G. 1999, MNRAS, 310, 1147 128, 129

Müller, W.-C., & Biskamp, D. 2000, Physical Review Letters, 84, 475 31, 51

Nagai, M., Tanaka, K., Kamegai, K., & Oka, T. 2007, PASJ, 59, 25 136

Nagayama, T., Omodaka, T., Handa, T., Iahak, H. B. H., Sawada, T., Miyaji, T., & Koyama, Y. 2007, PASJ, 59, 869 136

Nakamura, F., & Umemura, M. 2002, ApJ, 569, 549 77

Naoz, S., & Barkana, R. 2005, MNRAS, 362, 1047 60

BIBLIOGRAPHY

Natarajan, A., & Schwarz, D. J. 2008, Phys. Rev. D, 78, 103524 117, 133

Navarro, J. F., Frenk, C. S., & White, S. D. M. 1997, ApJ, 490, 493 112, 119, 128

Navarro, J. F., Ludlow, A., Springel, V., Wang, J., Vogelsberger, M., White, S. D. M., Jenkins, A., Frenk, C. S., & Helmi, A. 2008, ArXiv 0810.1522 129

Nolta, M. R., Dunkley, J., Hill, R. S., Hinshaw, G., Komatsu, E., Larson, D., Page, L., Spergel, D. N., Bennett, C. L., Gold, B., Jarosik, N., Odegard, N., Weiland, J. L., Wollack, E., Halpern, M., Kogut, A., Limon, M., Meyer, S. S., Tucker, G. S., & Wright, E. L. 2009, ApJS, 180, 296 20, 21, 46, 99

Norman, M. L., Bryan, G. L., Harkness, R., Bordner, J., Reynolds, D., O'Shea, B., & Wagner, R. 2007, ArXiv 0705.1556 69, 77

Núñez-López, R., Lipovka, A., & Avila-Reese, V. 2006, MNRAS, 369, 2005 94

O'Connell, R. F., & Matese, J. J. 1969, Nature, 222, 649 9

Oh, S. P. 2001, ApJ, 553, 499 27

Oh, S. P., & Haiman, Z. 2002, ApJ, 569, 558 29, 104, 112

Oh, S. P., & Mack, K. J. 2003, MNRAS, 346, 871 65, 168

Oka, T., Geballe, T. R., Goto, M., Usuda, T., & McCall, B. J. 2005, ApJ, 632, 882 136, 140

Olesen, P. 1997, Physics Letters B, 398, 321 31, 51

Olive, K. A. 2008, Advances in Space Research, 42, 581 33

O'Malley, T. F., Spruch, L., & Rosenberg, L. 1961, Journal of Mathematical Physics, 2, 491 179

Omont, A., Petitjean, P., Guilloteau, S., McMahon, R. G., Solomon, P. M., & Pécontal, E. 1996, Nature, 382, 428 135, 140

Omukai, K., Schneider, R., & Haiman, Z. 2008, ArXiv 0804.3141 7, 20, 28, 36, 46, 104, 173

O'Shea, B. W., Bryan, G., Bordner, J., Norman, M. L., Abel, T., Harkness, R., & Kritsuk, A. 2004, ArXiv Astrophysics e-prints 69, 77

O'Shea, B. W., & Norman, M. L. 2008, ApJ, 673, 14 102

Osterbrock, D. E. 1989, Astrophysics of gaseous nebulae and active galactic nuclei (Research supported by the University of California, John Simon Guggenheim Memorial Foundation, University of Minnesota, et al. Mill Valley, CA, University Science Books, 1989, 422 p.) 25, 102

Padmanabhan, N., & Finkbeiner, D. P. 2005, Phys. Rev. D, 72, 023508 22

Papadopoulos, P. P., Kovacs, A., Evans, A. S., & Barthel, P. 2008, A&A, 491, 483 137, 146

BIBLIOGRAPHY

Parker, L. 1968, Physical Review Letters, 21, 562 12

Pedersen, H. B. e. a. 2005, Phys. Rev. A, 72, 012712 76

Peebles, P. J. E. 1968, ApJ, 153, 1 4, 68, 72

—. 1993, Principles of physical cosmology (Princeton Series in Physics, Princeton, NJ: Princeton University Press, —c1993) 74, 153

Pentericci, L., Fan, X., Rix, H.-W., Strauss, M. A., Narayanan, V. K., Richards, G. T., Schneider, D. P., Krolik, J., Heckman, T., Brinkmann, J., Lamb, D. Q., & Szokoly, G. P. 2002, AJ, 123, 2151 1, 7, 171

Penzias, A. A., & Wilson, R. W. 1965, ApJ, 142, 419 4

Pequignot, D., Petitjean, P., & Boisson, C. 1991, A&A, 251, 680 24, 49, 73, 187

Pérez-Beaupuits, J. P., Aalto, S., & Gerebro, H. 2007, A&A, 476, 177 147

Pérez-Beaupuits, J. P., Spaans, M., van der Tak, F. F. S., Aalto, S., García-Burillo, S., Fuente, A., & Usero, A. 2009, A&A, 503, 459 147

Pierpaoli, E. 2004, Physical Review Letters, 92, 031301 22

Pinto, C., & Galli, D. 2008, A&A, 484, 17 30, 50, 181

Poelman, D. R., & Spaans, M. 2005, A&A, 440, 559 142

—. 2006, A&A, 453, 615 142

Press, W. H., & Schechter, P. 1974, ApJ, 187, 425 5, 26, 52

Price, D. J., & Bate, M. R. 2007, Ap&SS, 311, 75 174

—. 2008, MNRAS, 385, 1820 15

Pritchard, J. R., & Furlanetto, S. R. 2006, MNRAS, 367, 1057 56, 108

Purcell, E. M., & Field, G. B. 1956, ApJ, 124, 542 45

Puy, D., Alecian, G., Le Bourlot, J., Leorat, J., & Pineau Des Forets, G. 1993, A&A, 267, 337 4, 68, 101

Puy, D., & Signore, M. 2007, New Astronomy Review, 51, 411 68, 81, 82, 83

Quashnock, J. M., Loeb, A., & Spergel, D. N. 1989, ApJ, 344, L49 11

Ranalli, P., Comastri, A., & Setti, G. 2003, A&A, 399, 39 27

Rasera, Y., & Teyssier, R. 2006, A&A, 445, 1 119, 128

Ratra, B. 1992, ApJ, 391, L1 13

Ratra, B., & Peebles, P. J. E. 1995, Phys. Rev. D, 52, 1837 13

Rebhan, A. 1992, ApJ, 392, 385 10

Rees, M. J. 1987, QJRAS, 28, 197 6, 174

Ricotti, M. 2003, MNRAS, 344, 1237 129

Riechers, D. A., Walter, F., Bertoldi, F., Carilli, C. L., Aravena, M., Neri, R., Cox, P., Weiß, A., & Menten, K. M. 2009, ApJ, 703, 1338 135, 140, 149

Riechers, D. A., Walter, F., Brewer, B. J., Carilli, C. L., Lewis, G. F., Bertoldi, F., & Cox, P. 2008a, ApJ, 686, 851 135, 140

Riechers, D. A., Walter, F., Carilli, C. L., Bertoldi, F., & Momjian, E. 2008b, ApJ, 686, L9 135, 140

Righi, M., Hernández-Monteagudo, C., & Sunyaev, R. A. 2008a, A&A, 489, 489 95

—. 2008b, A&A, 478, 685 95

Riotto, A., & Trodden, M. 1999, Annual Review of Nuclear and Particle Science, 49, 35 3, 11

Ripamonti, E., Iocco, F., Ferrara, A., Schneider, R., Bressan, A., & Marigo, P. 2010, ArXiv e-prints 176

Ripamonti, E., Mapelli, M., & Ferrara, A. 2007a, MNRAS, 374, 1067 22, 33, 34, 35, 117, 133

—. 2007b, MNRAS, 375, 1399 22, 33, 117

Roberge, W., & Dalgarno, A. 1982, ApJ, 255, 489 90

Roos, M. 2003, Introduction to cosmology (Information and Computation) 3

Rubiño-Martín, J. A., Chluba, J., & Sunyaev, R. A. 2006, MNRAS, 371, 1939 68, 94

—. 2008, A&A, 485, 377 68, 94

Ruden, S. P., Glassgold, A. E., & Shu, F. H. 1990, ApJ, 361, 546 86, 87

Ruhl, J., Ade, P. A. R., Carlstrom, J. E., Cho, H.-M., Crawford, T., Dobbs, M., Greer, C. H., Halverson, N. w., Holzapfel, W. L., Lanting, T. M., Lee, A. T., Leitch, E. M., Leong, J., Lu, W., Lueker, M., Mehl, J., Meyer, S. S., Mohr, J. J., Padin, S., Plagge, T., Pryke, C., Runyan, M. C., Schwan, D., Sharp, M. K., Spieler, H., Staniszewski, Z., & Stark, A. A. 2004, in Presented at the Society of Photo-Optical Instrumentation Engineers (SPIE) Conference, Vol. 5498, Society of Photo-Optical Instrumentation Engineers (SPIE) Conference Series, ed. C. M. Bradford, P. A. R. Ade, J. E. Aguirre, J. J. Bock, M. Dragovan, L. Duband, L. Earle, J. Glenn, H. Matsuhara, B. J. Naylor, H. T. Nguyen, M. Yun, & J. Zmuidzinas, 11–29 95

Salucci, P., & Burkert, A. 2000, ApJ, 537, L9 129

Salvaterra, R., Haardt, F., & Ferrara, A. 2005, MNRAS, 362, L50 7, 20, 65, 110

BIBLIOGRAPHY

Sasaki, S., & Takahara, F. 1993, PASJ, 45, 655 24, 50, 68

Saslaw, W. C., & Zipoy, D. 1967, Nature, 216, 976 4, 68

Savin, D. W. 2002, ApJ, 566, 599 77

Savin, D. W., Krstić, P. S., Haiman, Z., & Stancil, P. C. 2004, ApJ, 607, L147 187

Scalo, J. 1998, in Astronomical Society of the Pacific Conference Series, Vol. 142, The Stellar Initial Mass Function (38th Herstmonceux Conference), ed. G. Gilmore & D. Howell, 201–+ 19, 29, 65, 104

Schaerer, D. 2002, A&A, 382, 28 5, 100

Schinnerer, E., Eckart, A., Tacconi, L. J., Genzel, R., & Downes, D. 2000, ApJ, 533, 850 148

Schleicher, D. R. G., Banerjee, R., & Klessen, R. S. 2008a, Phys. Rev. D, 79, 043510 33, 128, 130, 153

—. 2008b, Phys. Rev. D, 78, 083005 5, 7, 19, 46, 47, 50, 52, 53, 65, 99, 100, 101, 104, 106, 107, 119, 133, 153

—. 2009a, ApJ, 692, 236 21, 43, 110

Schleicher, D. R. G., Galli, D., Palla, F., Camenzind, M., Klessen, R. S., Bartelmann, M., & Glover, S. C. O. 2008c, A&A, 490, 521 43, 66, 101

Schleicher, D. R. G., Glover, S. C. O., Banerjee, R., & Klessen, R. S. 2009b, Phys. Rev. D, 79, 023515 41, 119

Schleicher, D. R. G., Spaans, M., & Klessen, R. S. 2008d, ArXiv 0812.3950 110

—. 2010, ArXiv e-prints 1001.2118

Schneider, R., Ferrara, A., Salvaterra, R., Omukai, K., & Bromm, V. 2003, Nature, 422, 869 6, 173

Schneider, R., Salvaterra, R., Choudhury, T. R., Ferrara, A., Burigana, C., & Popa, L. A. 2008, MNRAS, 384, 1525 7, 58

Schneider, R., Salvaterra, R., Ferrara, A., & Ciardi, B. 2006, MNRAS, 369, 825 20, 24, 50, 52, 101, 104

Scott, P., Fairbairn, M., & Edsjö, J. 2008a, ArXiv 0809.1871 98

—. 2008b, ArXiv 0810.5560 98

Seager, S., Sasselov, D. D., & Scott, D. 1999, ApJ, 523, L1 22, 23, 24, 47, 50, 68, 73, 76, 77, 81, 82, 101

—. 2000, ApJS, 128, 407 4, 22, 23, 24, 47, 50, 68, 73, 101

BIBLIOGRAPHY

Seljak, U., & Zaldarriaga, M. 1996, ApJ, 469, 437 73

Seshadri, T. R., & Subramanian, K. 2001, Physical Review Letters, 87, 101301 31, 50

Sethi, S. K. 2005, MNRAS, 363, 818 65, 168

Sethi, S. K., Nath, B. B., & Subramanian, K. 2008, MNRAS, 387, 1589 15, 30, 46

Sethi, S. K., & Subramanian, K. 2005, MNRAS, 356, 778 21, 26, 30, 31, 32, 46, 50, 51, 52, 64, 110

Shang, H., Glassgold, A. E., Shu, F. H., & Lizano, S. 2002, ApJ, 564, 853 30

Shankar, F., Salucci, P., Granato, G. L., De Zotti, G., & Danese, L. 2004, MNRAS, 354, 1020 139

Shankar, F., Weinberg, D. H., & Miralda-Escudé, J. 2009, ApJ, 690, 20 xvi, 139, 152, 154, 165

Shapiro, P. R., & Giroux, M. L. 1987, ApJ, 321, L107 7, 20, 50, 101

Shapiro, P. R., & Kang, H. 1987, ApJ, 318, 32 24

Shapiro, S. L. 2005, ApJ, 620, 59 152

Shchekinov, Y. A., & Vasiliev, E. O. 2007, MNRAS, 379, 1003 65, 169

Sheth, R. K., Mo, H. J., & Tormen, G. 2001, MNRAS, 323, 1 5

Shiromizu, T. 1998, Physics Letters B, 443, 127 31, 51

Shu, F. 1991, Physics of Astrophysics: Volume I Radiation (Published by University Science Books, 648 Broadway, Suite 902, New York, NY 10012, 1991.) 85

Shull, J. M., & van Steenberg, M. E. 1985, ApJ, 298, 268 27, 34, 58

Sigl, G., Olinto, A. V., & Jedamzik, K. 1997, Phys. Rev. D, 55, 4582 11, 172

Silk, J., & Langer, M. 2006, MNRAS, 371, 444 15, 20, 28, 46, 173

Sizun, P., Cassé, M., & Schanne, S. 2006, Phys. Rev. D, 74, 063514 14, 128, 131

Smith, B., Sigurdsson, S., & Abel, T. 2008a, MNRAS, 385, 1443 104

Smith, B. D., Turk, M. J., Sigurdsson, S., O'Shea, B. W., & Norman, M. L. 2008b, ArXiv e-prints, 806 104

Smith, F. J. 1966, Planet. Space Sci., 14, 929 46

Spaans, M., & Meijerink, R. 2008, ApJ, 678, L5 135, 142, 154, 156, 171

Spaans, M., & Silk, J. 2006, ApJ, 652, 902 6, 152

Spergel, D. N., Bean, R., Doré, O., Nolta, M. R., Bennett, C. L., Dunkley, J., Hinshaw, G., Jarosik, N., Komatsu, E., Page, L., Peiris, H. V., Verde, L., Halpern, M., Hill, R. S., Kogut,

BIBLIOGRAPHY

A., Limon, M., Meyer, S. S., Odegard, N., Tucker, G. S., Weiland, J. L., Wollack, E., & Wright, E. L. 2007, ApJS, 170, 377 67, 82

Spergel, D. N., & Press, W. H. 1985, ApJ, 294, 663 116, 117

Spinoglio, L., Malkan, M. A., Smith, H. A., González-Alfonso, E., & Fischer, J. 2005, ApJ, 623, 123 141, 142

Spolyar, D., Freese, K., & Gondolo, P. 2008, Physical Review Letters, 100, 051101 16, 19, 20, 33, 35, 98, 99, 110, 153

Spoon, H. W. W., Marshall, J. A., Houck, J. R., Elitzur, M., Hao, L., Armus, L., Brandl, B. R., & Charmandaris, V. 2007, ApJ, 654, L49 147

Springel, V., Wang, J., Vogelsberger, M., Ludlow, A., Jenkins, A., Helmi, A., Navarro, J. F., Frenk, C. S., & White, S. D. M. 2008a, ArXiv 0809.0898 129

Springel, V., White, S. D. M., Frenk, C. S., Navarro, J. F., Jenkins, A., Vogelsberger, M., Wang, J., Ludlow, A., & Helmi, A. 2008b, ArXiv 0809.0894 129

Stancil, P. C., Lepp, S., & Dalgarno, A. 1998, ApJ, 509, 1 4, 68, 83, 101, 183, 187

Starobinskij, A. A. 1982, Physics Letters B, 117, 175 2

Stecher, T. P., & Williams, D. A. 1967, ApJ, 149, L29+ 6, 76

Steidel, C. C., Pettini, M., & Adelberger, K. L. 2001, ApJ, 546, 665 29, 105

Strong, A. W., Moskalenko, I. V., & Reimer, O. 2004, ApJ, 613, 956 xii, xiii, 111, 113, 114, 115, 131

Subramanian, K. 2000, ApJ, 538, 517 129

Subramanian, K., & Barrow, J. D. 1998, Phys. Rev. D, 58, 083502 31, 50, 52

Sunyaev, R. A., & Zeldovich, Y. B. 1972, A&A, 20, 189 74

Switzer, E. R., & Hirata, C. M. 2008a, Phys. Rev. D, 77, 083006 68

—. 2008b, Phys. Rev. D, 77, 083008 68

Tan, J. C., & Blackman, E. G. 2004, ApJ, 603, 401 15

Taoso, M., Bertone, G., Meynet, G., & Ekstrom, S. 2008, ArXiv 0806.2681 16, 20, 98, 100, 106, 116

Tashiro, H., & Sugiyama, N. 2006, MNRAS, 368, 965 21, 26, 46, 64

Tashiro, H., Sugiyama, N., & Banerjee, R. 2006, Phys. Rev. D, 73, 023002 21, 45, 46, 110

Tasitsiomi, A., Kravtsov, A. V., Gottlöber, S., & Klypin, A. A. 2004, ApJ, 607, 125 129

Taylor, J. E., & Navarro, J. F. 2001, ApJ, 563, 483 129

Tegmark, M., Silk, J., Rees, M. J., Blanchard, A., Abel, T., & Palla, F. 1997, ApJ, 474, 1 5

BIBLIOGRAPHY

Thompson, T. A., Quataert, E., & Murray, N. 2005, ApJ, 630, 167 141, 177

Tornatore, L., Ferrara, A., & Schneider, R. 2007, MNRAS, 382, 945 36, 173

Treister, E., Urry, C. M., & Virani, S. 2009, ArXiv 0902.0608 xvi, 154, 165

Trevisan, C. S., & Tennyson, J. 2002, Plasma Physics and Controlled Fusion, 44, 1263 187

Turner, M. S., & Widrow, L. M. 1988, Phys. Rev. D, 37, 2743 12, 13

Ullio, P., Bergström, L., Edsjö, J., & Lacey, C. 2002, Phys. Rev. D, 66, 123502 110

Vainshtein, S. I., Zeldovich, I. B., & Ruzmaikin, A. A. 1980, Moscow Izdatel Nauka 8

Valdés, M., Ferrara, A., Mapelli, M., & Ripamonti, E. 2007, MNRAS, 377, 245 117

Verner, D. A., & Yakovlev, D. G. 1995, A&AS, 109, 125 139

Wada, K., & Norman, C. A. 2007, ApJ, 660, 276 136

Wada, K., Papadopoulos, P. P., & Spaans, M. 2009, ApJ, 702, 63 140, 149

Walter, F., Carilli, C., Bertoldi, F., Menten, K., Cox, P., Lo, K. Y., Fan, X., & Strauss, M. A. 2004, ApJ, 615, L17 135, 140

Walter, F., Riechers, D., Cox, P., Neri, R., Carilli, C., Bertoldi, F., Weiss, A., & Maiolino, R. 2009a, ArXiv e-prints 135, 140, 149

Walter, F., Riechers, D. A., Carilli, C. L., Bertoldi, F., Weiss, A., & Cox, P. 2007, in Astronomical Society of the Pacific Conference Series, Vol. 375, From Z-Machines to ALMA: (Sub)Millimeter Spectroscopy of Galaxies, ed. A. J. Baker, J. Glenn, A. I. Harris, J. G. Mangum, & M. S. Yun, 182–+ 135, 140

Walter, F., Weiß, A., Riechers, D. A., Carilli, C. L., Bertoldi, F., Cox, P., & Menten, K. M. 2009b, ApJ, 691, L1 142

Watanabe, K., Hartmann, D. H., Leising, M. D., & The, L.-S. 1999, ApJ, 516, 285 xii, xiii, 120, 121, 131

Weidenspointner, G., Shrader, C. R., Knödlseder, J., Jean, P., Lonjou, V., Guessoum, N., Diehl, R., Gillard, W., Harris, M. J., Skinner, G. K., von Ballmoos, P., Vedrenne, G., Roques, J.-P., Schanne, S., Sizun, P., Teegarden, B. J., Schönfelder, V., & Winkler, C. 2006, A&A, 450, 1013 14, 97, 127

Weiß, A., Downes, D., Neri, R., Walter, F., Henkel, C., Wilner, D. J., Wagg, J., & Wiklind, T. 2007, A&A, 467, 955 135, 136, 148

Weiß, A., Downes, D., Walter, F., & Henkel, C. 2005, A&A, 440, L45 135, 140

Whalen, D., Abel, T., & Norman, M. L. 2004, ApJ, 610, 14 28, 53, 104

Widrow, L. M. 2002, Reviews of Modern Physics, 74, 775 21

BIBLIOGRAPHY

Wise, J. H., & Abel, T. 2007a, ApJ, 665, 899 6, 176

—. 2007b, ApJ, 671, 1559 102

—. 2008, ApJ, 684, 1 53, 103

Wishart, A. W. 1979, MNRAS, 187, 59P 86

Wong, W. Y., Moss, A., & Scott, D. 2008, MNRAS, 386, 1023 23, 47, 68

Wood, K., & Loeb, A. 2000, ApJ, 545, 86 28, 104

Wouthuysen, S. A. 1952, AJ, 57, 31 47, 95, 108

Wyithe, J. S. B., & Loeb, A. 2003, ApJ, 588, L69 5

Xu, H., O'Shea, B. W., Collins, D. C., Norman, M. L., Li, H., & Li, S. 2008, ApJ, 688, L57 43, 174

Yamazaki, D. G., Ichiki, K., Kajino, T., & Mathews, G. J. 2006, ApJ, 646, 719 9

—. 2008, Phys. Rev. D, 78, 123001 9

Yoon, S.-C., Iocco, F., & Akiyama, S. 2008, ArXiv 0806.2662 xviii, 16, 20, 33, 98, 100, 101, 106, 107, 110, 115, 116, 123, 168

Yoshida, N., Abel, T., Hernquist, L., & Sugiyama, N. 2003, ApJ, 592, 645 27

Yoshida, N., Oh, S. P., Kitayama, T., & Hernquist, L. 2007a, ApJ, 663, 687 6, 20, 36, 104

Yoshida, N., Omukai, K., & Hernquist, L. 2007b, ApJ, 667, L117 6, 20, 36, 104

Yoshida, N., Omukai, K., Hernquist, L., & Abel, T. 2006, ApJ, 652, 6 5, 76

Yüksel, H., Horiuchi, S., Beacom, J. F., & Ando, S. 2007, Phys. Rev. D, 76, 123506 110, 113, 128, 129, 133

Yüksel, H., Kistler, M. D., Beacom, J. F., & Hopkins, A. M. 2008, ApJ, 683, L5 20, 105

Zaldarriaga, M. 1997, Phys. Rev. D, 55, 1822 7

Zaldarriaga, M., Furlanetto, S. R., & Hernquist, L. 2004, ApJ, 608, 622 65, 168

Zeldovich, I. B., & Novikov, I. D. 1983, Relativistic astrophysics. Volume 2 - The structure and evolution of the universe /Revised and enlarged edition/ (Chicago, IL, University of Chicago Press, 1983, 751 p. Translation.) 21

Zeldovich, I. B., Ruzmaikin, A. A., Sokolov, D. D., & Turchaninov, V. I. 1979, Moscow Institut Prikladnoi Matematiki AN SSSR 8

Zeldovich, Y. B., Kurt, V. G., & Syunyaev, R. A. 1969, Soviet Journal of Experimental and Theoretical Physics, 28, 146 4, 68

Zhang, L., Chen, X., Kamionkowski, M., Si, Z.-G., & Zheng, Z. 2007, Phys. Rev. D, 76, 061301 35, 43

BIBLIOGRAPHY

Zhang, L., Chen, X., Lei, Y.-A., & Si, Z.-G. 2006, Phys. Rev. D, 74, 103519 34, 43

Ziegler, U., & Rüdiger, G. 2000, A&A, 356, 1141 174

Zweibel, E. G. 2006, Astronomische Nachrichten, 327, 505 8

Zweibel, E. G., & Heiles, C. 1997, Nature, 385, 131 8

Zygelman, B. 2005, ApJ, 622, 1356 46, 56

Zygelman, B., Dalgarno, A., Kimura, M., & Lane, N. F. 1989, Phys. Rev. A, 40, 2340 187

Zygelman, B., Stancil, P. C., & Dalgarno, A. 1998, ApJ, 508, 151 88, 89, 187

VDM Verlagsservicegesellschaft mbH

Die VDM Verlagsservicegesellschaft sucht für wissenschaftliche Verlage abgeschlossene und herausragende

Dissertationen, Habilitationen, Diplomarbeiten, Master Theses, Magisterarbeiten usw.

für die kostenlose Publikation als Fachbuch.

Sie verfügen über eine Arbeit, die hohen inhaltlichen und formalen Ansprüchen genügt, und haben Interesse an einer honorarvergüteten Publikation?

Dann senden Sie bitte erste Informationen über sich und Ihre Arbeit per Email an *info@vdm-vsg.de*.

Sie erhalten kurzfristig unser Feedback!

VDM Verlagsservicegesellschaft mbH
Dudweiler Landstr. 99 Telefon +49 681 3720 174
D - 66123 Saarbrücken Fax +49 681 3720 1749
www.vdm-vsg.de

Die VDM Verlagsservicegesellschaft mbH vertritt

Printed by Books on Demand GmbH, Norderstedt / Germany